UNLEASHING THE POWER OF

3P

The Key to Breakthrough Improvement

UNLEASHING THE POWER OF

3P

Dan
McDonnell

Drew A.
Locher

The Key to Breakthrough Improvement

CRC Press
Taylor & Francis Group
Boca Raton London New York

CRC Press is an imprint of the
Taylor & Francis Group, an **informa** business

A PRODUCTIVITY PRESS BOOK

CRC Press
Taylor & Francis Group
6000 Broken Sound Parkway NW, Suite 300
Boca Raton, FL 33487-2742

© 2013 by Taylor & Francis Group, LLC
CRC Press is an imprint of Taylor & Francis Group, an Informa business

No claim to original U.S. Government works

Version Date: 2012912

ISBN 13: 978-1-4398-8612-0 (pbk)

Library of Congress Cataloging-in-Publication Data

McDonnell, Dan.
    Unleashing the power of 3P : the key to breakthrough improvement / Dan McDonnell, Drew A. Locher.
        p. cm.
    Includes bibliographical references and index.
    ISBN 978-1-4398-8612-0 (hbk. : alk. paper)
    1. Production management. 2. Industrial productivity. 3. Lean manufacturing. I. Locher, Drew. II. Title.

TS155.M3456 2013
658.5--dc23   2012035047

Visit the Taylor & Francis Web site at
http://www.taylorandfrancis.com

and the CRC Press Web site at
http://www.crcpress.com

For my wife Virginia and my daughters Amanda and Megan, who have always supported me in all that I have taken on. My achievements are theirs.

—Dan McDonnell

This book is dedicated to my entire extended family: my wife Eileen, parents Adele and Walt, aunt and uncle and "second parents" Dot and Har, my sister Lynn whom I adore, her husband Steve, my brother Scott whom I admire, his wife Janice, and my nieces and nephews Bryan, Jonathan, Janine, Emily, Patrick, Grace, Pierce, Nina, and Casey.

—Drew Locher

Contents

Unleashing the Power of 3P

Character Names, Roles, and Other Information

Corporate Name	Enterride	
Division	Utility Division	
Division Name	Trail Rider	Memphis Operation
Consulting Company	Kata Movements	
Old Product Families	"Trail Rider" and "Traction"	
Acquired Product Family	"Trail Gripper"	
Frank Kent	Corporate	Chief Executive Officer (CEO)
Steve Sawyer	Corporate	Vice Chair
John Cuso	Corporate	Chief Financial Officer (CFO)
Paula Angle	Corporate	Senior Vice President, Human Resources
Pete Grant	Trail Rider	Plant Manager
Dave Martin	Trail Rider	Lean Leader
Mike Young	Trail Rider	Shop Manager
George Hall	Trail Rider	Advanced Manufacturing Engineering Leader
John Lee	Trail Rider	Lean Leader
Gina Nelson	Trail Rider	Lean Leader
Mary Long	Trail Rider	Inside Sales Manager
Sylvia Bennett	Trail Rider	Order entry

Bill Stark	Trail Rider	Material Manager
Lou Marks	Trail Rider	Quality Manager
Johnny Cox	Trail Rider	Shop Assembler
Gerry Barr	Trail Rider	Engine Assembler
Bill Cook	Trail Rider	Maintenance/Moonshiner
Norm Wilson	Trail Rider	Maintenance/Moonshiner
Byron Hill	Trail Rider	Maintenance/Moonshiner
Harry Givens	Trail Rider	Maintenance Manager
Brenda Lewis	Trail Rider	Product Design Engineer
Ben Smith	Trail Rider	Product Design Engineer
Betty King	Trail Rider	Metals Commodity Leader
Scott Green	Trail Rider	Material Controller
Linda Campbell	Trail Rider	Supplier Quality Engineer
Rick	Trail Rider	Mail deliverer
John	Trail Rider	Mail deliverer
Blake Derry	Trail Rider	Marketing Manager
Ken Saguchi	Kata Movements	Sensei
Joan Jarrett	St. Lucia's Hospital	CEO
Megan Taylor	St. Lucia's Hospital	Doctor
Virginia Hall	St. Lucia's Hospital	Nursing Supervisor
Amanda Stewart	St. Lucia's Hospital	Nurse
Debra Grant	Family	Pete Grant's wife
Julie Martin	Family	Dave Martin's wife
Tess Martin	Family	Dave Martin's daughter
Eileen Saguchi	Family	Ken Saguchi's wife

List of Exhibits

List of Tables

Preface

"Close them!" The Enterride CFO's words meant impending doom for his plant. Recent improvements at Enterride's Trail Rider Memphis Operation had bought them a little more time, but Pete Grant, the Trail Rider Plant Manager, knew much more was needed to save it. He needed a real breakthrough, a true game changer. But what? How? They had been applying all the Lean concepts he was aware of. Was there something more? Someone told him about a process called "3P." Could this be just the thing he was looking for? Pete didn't know much about it, but he did know that it required a significant investment of people's time early in a product and process development project. Could he afford that investment? It was a big risk, but one he had to take if he hoped to save his plant.

In the following pages, we'll follow Pete and his team's journey through Lean and 3P from product acquisition to product launch, as they learn how to trim waste, increase efficiency, and bring their plant back from the brink. We'll tag along with them as they visit a hospital and learn how Lean and 3P can be applied to *any* industry. We'll celebrate their successes and learn from their mistakes.

Lean and 3P can transform your organization. Let Enterride's journey show you just how transformative they can be.

Acknowledgments

My Lean journey started in 1988 with two special words: "Focused Factory." My Lean flame was lit and has continued to glow brighter and brighter ever since. There are so many folks who have helped me learn and grow throughout my career that I could likely write a full book of acknowledgments. It is literally on the backs of thousands of wonderful "coaches" that I achieved enough learning to help me, as a continuing "student," to write this book.

I would like to share the prime reason that I took the time to capture my experiences. I have learned most of what I know from reaching out to others, and it was through their willing generosity that a deep spirit of responsibility to share back was born within me, along with an enduring purpose to help advance excellence within our wealth-generating manufacturing community. I have reflected upon the specific individuals who perhaps had the greatest influence on my 3P journey and felt it was most appropriate to acknowledge them individually.

Tom Bechtel, my first Lean teacher

Norris Woodruff, who taught me the power of simulation and proactive work

Dave Hogg, who became a lifetime friend and my most trusted and valued Lean mentor

Todd Wyman, my friend and colleague, who reached out to me twice to join him in pursuit of major transformations, convinced me I had a different calling in the working world, and has believed in and supported me completely

The Shingijutsu Corporation became my prime mentor, and while I worked under the observation of many great "Shingi" senseis, I have to especially thank Mr. Sato, Mr. Nakai, and mostly Mr. Nagamatsu, who had the vision and courage to develop a new 3P teaching and coaching process for us

Steve Pillsbury of PRTM, who taught me much more about material flow

Those 3P pioneers at GE Appliances who helped me further benchmark and learn

Mike Lamach, who has pushed me to a whole new level of coaching capability

Keith Sultana, who has become IR's first 3P pioneer

Drew Locher, who approached me to co-write this book with him

Finally, through all of this, through many hours and steadfast dedication in my working life, it is the balance in my personal life that has kept me grounded and nurtured. I can only thank my wife Virginia, and my children Amanda and Megan, for being the backbone of my life, for all of their support, and for their belief in me. Perhaps, ultimately, they have been the real reason I have been doing all of this. For in the end, what really matters in life is family and friends; they are undoubtedly the most important parts of life.

To all of those who have helped me, thank you. Thank you from the bottom of my heart. May your Lean torch and your interest in 3P continue to burn on with vigor.

—Dan McDonnell

As I reflect on the past and decide who to acknowledge for contributions to this book, as well as my career as an agent for change in business, I am compelled to acknowledge the General Electric Company, which introduced me to the concepts of "Integrated Product and Process Development" in the 1980s. I also want to acknowledge Ken Rolfes for twenty years of coaching, not just in this subject but in many others, as well as my wife Eileen Locher and John Toussaint for their contributions to the healthcare content.

I would like to thank Michael Sinocchi of Productivity Press for the opportunity to continue to contribute to "gaps" in the body of knowledge in the broad subject of Operational Excellence. Many thanks to our editor Kirsten Miller, who always makes the editing process a collaborative and enjoyable one, as well as Kevin Mercer, who helps us bring the words to life with his graphics. Most importantly, I want to thank my co-author Dan McDonnell for giving me an opportunity to help him bring his story to the world.

—Drew Locher

About the Authors

Dan McDonnell began his career at Multilin, a small high-tech firm in Canada, where he rose through myriad manufacturing assignments to the level of vice president of Operations, and where he started his Lean journey. He spent 14 years with the General Electric Company, where he served as a plant manager, and eventually as a manufacturing general manager for 11 different factories. He most recently served for 3 years as the manager of the Lean Initiative for GE Transportation where his deep experience with 3P really began. McDonnell is currently the vice president of Operational Excellence for Ingersoll Rand.

Along with his professional career, Dan developed some of his Lean expertise through volunteer efforts and networking within the Association for Manufacturing Excellence (AME), serving 3 years as Canadian Regional President, 7 years on the national board, and a term as national president. He has also chaired two national Lean conferences, run marketing for three others, and presented at numerous workshops and conferences.

McDonnell has been married for 29 years, has two children, one of whom is a registered nurse, and another who is attending medical school. He resides in Mooresville, NC.

Drew Locher first became involved in the development and delivery of innovative Business Improvement programs while working for General Electric in the 1980s. Since leaving GE in 1990 and forming Change Management Associates (CMA), he has provided Operational Excellence and organizational development services to industrial and service organizations representing a wide variety of industries including healthcare, transportation, distribution, education, financial services and manufacturing.

In the late 1990s, Locher helped the National Institute of Standards and Technologies' Manufacturing Extension Partnership (MEP) to develop a Lean university. Since 2001, he has been a faculty member of the Lean Enterprise Institute (LEI), the not-for-profit organization of the co-author of "Lean Thinking." Locher is also an adjunct faculty member at the Fisher College of Business at The Ohio State University.

In 2004, Locher co-authored a book titled, *The Complete Lean Enterprise: Value Stream Mapping for Administrative and Office Processes*. The book received a 2005 Shingo Prize for Research. In 2008 he published, *Value Stream Mapping for Lean Development*. His third book titled, *Lean Office and Service Simplified: The Definitive How-to Guide* was released in 2011 and is a 2012 Shingo Prize recipient.

Chapter 1

A Case for Change

"Close them!"

The words, spoken forcefully and in anger, came from the far side of the table, where harsh rays of light beamed through the corner windows, exposing bands of floating dust pellets. John Cuso, CFO of Enterride, was a man of few words, but he'd finally had enough; perhaps this irritating dust cloud around him somehow represented what he felt of that Trail Rider division he'd come to despise. Seated around the table with him were Chairman and CEO Frank Kent, Vice Chairman Steve Sawyer, and Senior Vice President for Human Resources Paula Angle.

The launch of Trail Rider, the new vehicle in their Utility Division, had been poorly handled. Customers had staged a revolt. They'd lost their biggest customer to a competitor, and other customers were also threatening to leave. The Memphis operation continued to lose money.

"I thought you said Pete Grant could fix this business," John lashed out at Steve. He hadn't gotten over Steve's decision to promote Pete Grant to plant manager of the Memphis operation—without John's approval.

Steve, in a tone that matched John's, responded, "Pete Grant has headed that division for just six months, "and we didn't give him much to work with. He needs more time."

"Everyone calm down. Let's think this through," Paula chimed in.

They all knew that the CEO had been favoring looking for a buyer and divesting their Utility Division. After six years of trying to expand markets outside of North America and increase operating income, the division remained a drag on total margin rates, and market analysts were calling for a breakup of the company.

John Cuso tried again to make his case to close the Memphis plant and move the operation outside the United States. He and his team were working on a report showing that a relocation to Monterrey, Mexico, would result in significantly lower labor costs. He asked for another week to finish the report, promising that the financial benefits of a close-and-relocate versus a sale would become clear. He had been on the original acquisition team that purchased the Utility Division six years ago, and it was still one of his favored children.

But Steve was having none of it. They were going to turn around the Memphis business.

> "Pete is getting a handle on the division. I like the changes he's making. Besides, he's just hired a new Lean Leader from a Toyota affiliate, and the two of them have committed to bringing the customer quality issues under control and returning to profitability by the end of the year. I have faith in Pete, and I say we should give them some more time."

Frank listened thoughtfully and told Steve he had until the end of the year to right the wrongs, calm customer noise, and turn a profit in Memphis. Otherwise, the Division would be up for sale or be moved to Mexico.

Steve reiterated his commitment to do what it took to make Memphis work. John sat glumly, staring into space. Paula sighed. She could tell that they hadn't heard the end of it. Steve and John still were far from seeing eye to eye on many things, and there was no way that the lid would remain on this one.

※　※　※

Dave Martin woke to slants of light easing their way through the wooden blinds of his bedroom. A cool breeze gusted through the open window. He heard stifled giggles in the hall before the door creaked slowly open, and sensed his wife and children softly approaching his side of the bed. He heard a chorus of "Happy Birthday, Dad!" and felt himself being shaken.

Thirty-eight was going to be a good year. Last week's Lean Enterprise conference had emboldened him, energized him, and fed his determination to approach his new team and sell a clear need for change. Despite yesterday's grueling day of travel and a midnight return home, he willed himself up out of bed for a homemade breakfast, after a minute or two of gentle prodding from his youngest daughter Tess.

Sitting on the patio beside Julie, his wife of 15 years, and looking out over his kids running in the backyard, he thought about the ups and downs of a great career to date. He was convinced that his big break was just around the corner. His old friend Pete Grant had been wooing him for the past two years, finally convincing him to join Trail Rider as a much-needed facilitator and coach. They both knew that this was an unprecedented opportunity to drive a major transformation.

Dave didn't know what he was walking into, but it was probably just as well. The weekend before his first big day had been great.

❋   ❋   ❋

Pete Grant, Trail Rider's plant manager, stood in the receiving bay with his material manager that Saturday morning, trying to understand what had happened to the control modules that were supposed to be in Friday night to facilitate a desperately needed productive weekend of output. Fifty operators, on overtime, were standing by—doing busy work, or drinking coffee and talking about last night's big playoff game. It hadn't been the first time their biggest, and most problematic, supplier had let him down. But he had bigger problems on his mind after the call he'd received first thing this morning. It was one of the regional sales managers reporting that they'd just lost their biggest customer, about 10% of their total volume, to their main competitor. With the year they had had to date, they didn't need this new headache. Pete had always been a confident, positive, upbeat guy, but after six months, even he was starting to doubt his ability. It was an uncomfortable and unusual feeling and he hoped not a bad omen for the future.

❋   ❋   ❋

Sitting on a business-class flight, returning to the United States from a two-week training program in Japan, Ken Saguchi was unwinding, clicking through the movie offerings on the lousy airplane television screens. He turned off the television and reflected on a telephone conversation that he'd had with his mother months earlier. "Your dad has been having these headaches," his mother had said one night on the phone. And now he was gone, succumbing to brain cancer just four months after that phone call. The loss seriously rattled him. Besides the understandable deep sadness, he was feeling uncertain about himself, was impatient with others, and had developed a quick temper.

Three months prior to that call, he'd had a long conversation with his dad. After 20 years of the best business and manufacturing learning a person could be paid to do at Toyota, with a great pension and the remainder of his career ahead of him, several former colleagues had been wooing him to join their Lean consulting company.

For his entire life, his dad had been his most trusted advisor. He had long told him that the key to career happiness is to find something that you're both passionate about *and* good at. It was great advice. He thought back to a conversation he'd had with his father just before he'd left Toyota.

"What's your heart and mind telling you?"

"I don't know, Dad. I'm torn. Toyota has always treated me very well. I love the company and the opportunities I've had there. I was able to rise through the engineering ranks, and I learned something new every day. I'm still learning. What's scariest, in some ways, is that I'm learning and growing now more than ever. Still, even though I'm far from an expert, I think I have the knowledge and skills that I can use to help other companies. I've seen so many companies shut down operations and move them to other countries in search of greater profits, or go out of business altogether. People losing their jobs; entire communities suffering. Most of it's unnecessary or just flat-out wrong. I know that I can help them with what I learned at Toyota. But Brent and Amy have college and marriage ahead, and retirement's creeping closer. I'm afraid of the risk."

"I can't tell you what to do, but I know that whatever you choose to do, you'll be successful and financially sound. I've known that since you saved every penny from your first paper route. I think you know what you want to do."

And he had known. As Ken embarked on his new career path, he'd had his family, especially his dad, for strength. But now, for the first time, he had to contemplate going on without him, and he was having a hard time.

For the four months his dad was sick, he had become very disillusioned about the state of the healthcare system. Toyota had taught him about *process* and how to take waste out of it. Although absolutely wonderful people were taking care of his father, he was astounded at the level of waste and inefficiency he observed. Ironically, his first assignment at Kata Movements, his new Lean consulting company, was to help St. Mary's Hospital Group with their first Lean transformation journey. At work, he could focus and lose himself. At home, he was still despondent over the loss of his father. He and his wife Eileen were fighting; and prior to leaving for Japan, they'd considered going to counseling.

And so he sat on the plane on the long flight home, ready to wrap up his work at St. Mary's and begin work with this new client, Trail Rider, in two weeks. Although he was excited about the new business opportunity, he was saddened by the uncertainty in his personal life.

❋　❋　❋

When the candles had been blown out and the ice cream put back in the freezer, Dave helped Julie load the dishwasher, looking forward to all the positive changes he knew were ahead.

❋　❋　❋

Pete had returned home from another Saturday at work. He sat in his home office, shades drawn, worried about his ability to fix a broken business for the first time in his career. He was comforted that his friend Dave was joining him, and was happy that Kata Movements, the consulting company he had just retained, was coming on board in two weeks. Maybe positive change was finally around the corner. For the first time in two months, he started to feel hopeful.

※   ※   ※

The CFO was tying up his boat, still steaming over the Trail Rider business and wanting them relocated or sold. The vice chair, putting his weed eater away in his garage, was thankful for the six-month reprieve his boss gave the team, but he knew the pressure for significant positive change was still coming around the corner. He hadn't heard the last of John Cuso. Sometimes he tired of the never-ending corporate political dance.

※   ※   ※

The CEO stretched out on his lounge chair, only half-watching the ball game, knowing the next six months were critical. His company had a lot at stake in restoring the Utility Division to profitability. The analysts, who for years had bet with him, were becoming more impatient. He knew he had limited time and needed to enter next year with a strong recovery underway. But he knew that the Utility Division was not his only issue. Inefficient processes were rife across his entire business, but he had not wanted to admit it. He'd been coached to spend more time out in the departments, and he couldn't ignore what his observations were telling him. Facts were facts. And the recent report from his audit staff had scared him. The company had a poor track record for product launches across the enterprise, rarely hitting launch dates and other targets. Customers were complaining mightily about quality and delivery levels. He just knew there had to be a better way. He needed real breakthrough change.

If there ever was a case for a big restructuring, this was it. But he knew that making big changes to Trail Riders' internal processes, and in other divisions and operations of the corporation, was the best way to right the ship.

In the end, he had faith in his team to fix it!!!

Chapter 2

Pre-3P: The Art of Air Guitar

Dave walked into Pete's office. Pete jumped up from behind his desk, not expecting him this early.

"I thought you weren't coming in until 8:00."
"Just wanted to get an early start," Dave said, as they shook hands.
"No time like the present," Pete said. "Let's walk the factory and show you around."

They went out to the factory and looked out over several long assembly lines and countless storage areas where semi-finished products had accumulated. There was also a shipping-and-receiving area, which appeared to be in disarray, and a large stockroom where large quantities of parts from suppliers were stored. During the walk, they observed many of the assembly processes, and this helped Dave get acclimated to the products that the factory produced. Pete introduced Dave to many people along the way. Dave was astounded at the lack of evident standard work, flow, and process discipline. And, as he'd seen previously in his career at other manufacturing operations, waste abounded everywhere. Each of these problems presented an opportunity for improvement. They stopped at a new section of the line, for the "Traction" product they'd just launched. Pete explained that they'd underestimated capability at launch, had quickly realized that they had real capacity issues, and had to expand the stations on the final assembly line.

A few team members were waiting in this area for them to get there. Pete introduced Dave to Mike Young, the first-shift shop manager, and George Hall, the advanced manufacturing engineering (AME) leader for the plant, both of whom had been involved in this line expansion effort. He also introduced Dave to two members who would report directly to him: the Lean leaders who'd helped develop this expansion, John Lee and Gina Nelson.

Pete said, "John and Gina, let me introduce you to your new boss, Dave Martin."

After an exchange of pleasantries, Dave looked at the four new stations they were adding. It was obvious to his trained eye that they'd used minimal data in determining how to increase capacity. There was a lot missing.

> Dave looked at Pete and asked, "How many weeks until you make these changes?"
>
> Pete said, "We're doing our first real trial this afternoon. We expect to formally switch over by Wednesday."
>
> Dave raised an eyebrow, but nobody noticed. He asked the team assembled around the first new station, "What do you see?"

The team looked puzzled. It was a simple station, and the station wasn't currently running. George, the AME leader, tried to explain that the station was where the frames would be assembled.

> Dave tried a different question, "What don't you see?"
>
> "What do you mean?" George asked.
>
> "Take a look. Tell me what you *don't* see."
>
> Gina Nelson, one of the Lean leaders, said, "I don't see a work-in-process (WIP) unit at the station."
>
> Pete said, "I don't see an operator, since we're not working it yet."
>
> Dave pressed for more. "What else?"

There were a few strange looks and headshakes among the group Mike, George, John, and Gina were wondering who this new guy was. There *were* no other answers.

> Dave said, "Let's try to build a unit here."
>
> Pete responded, "We don't have the frame or most of the parts here yet. The team is scrambling to get them ready by this afternoon."
>
> Dave said, "It doesn't matter; we don't need them to build the assembly."

George Hall's thoughts were written on his face. "This guy's an idiot." The rest of the team members were wearing similar expressions.

> Dave could see what they were thinking. "Let me build one—you guys watch."
>
> He climbed up on the platform, clapped his hands together, smiled enthusiastically, and said, "Let's do this. So what's first?"
>
> George said, "Well, you need a frame."
>
> "Great! Where are those kept?"
>
> George pointed. "They'll be kept over there, across the aisle."
>
> "Perfect," said Dave, as he cupped his hand over his eyes, and for effect, said "way over there?"
>
> Pete said, "We don't have everything in place. I don't understand how we can do this."

His friend said, "You ever play air guitar, Pete?"

"Sure. We all did, when we were kids."

"Exactly, and the same principles can be applied to work, simply and easily,
     with great results. Dave continued. "So what am I supposed to do first again?"

George said, "You need to get a frame."

"I'm supposed to do that?" Dave asked. "Okay, here goes."

Dave walked down off the platform and labored his way the 80 steps to
where George was pointing, where the frames would be stored, exaggerating
each step. The rest of the team followed behind. A few hourly operators downstream glanced over, curious about the procession, wondering what the heck was
going on in their area.

When he got to the frame storage space, Dave asked, "By the way, how am I
     supposed to know I'm starting with the frame and where to get it?"

George stepped forward. "My team's still putting together the work instruc-
     tions with the Quality Assurance folks. It'll take about 20 pages to document
     every move and all of the quality and safety steps for this workstation.
     They're going to be great, the best we've ever had, but we won't have them
     for another week."

Dave said mildly, "But we have a pilot this afternoon and need to start produc-
     tion Wednesday. How are the operators on all the shifts going to know what
     to do?"

Dave paused, but there was no response. "Also, this all sounds great. We're
     going to have a lot of new operators on these stations and they'll need some
     good training and reference material to get up to speed, but let me ask
     you a question. Does anybody think that the operators are going to use a
     20-page set of work instructions every time they work on a unit? Probably
     not! I wonder what else we might use for documentation that they might
     more easily use and want to use on a regular basis."

He let the sentence hang in the air for a second or two. He didn't want to give
the team all of the answers. His goal was to make them think. He could see George
seething. Dave didn't understand why this guy had such a huge beef with him, but
he couldn't coddle everyone for fear of offending George. He was determined to
win him over in time. He'd converted a lot harder cases than George.

"So how will the operators know what to do this afternoon?"

"A trainer will be with them," said George.

"And how will the trainer know what to do?"

George said abruptly, "They've been building off-road vehicles for more than
     25 years."

"Okay, but have they ever built them at this particular station?" Dave said rhe-
     torically. "Well, of course not. We've never had this station here before."

The two other Lean leaders looked at each other. They were starting to like their new boss and could tell that this was going to be interesting.

Pete was already beginning to suspect that the new station design was a disaster. He could see that George was getting very irritated and becoming more and more defensive.

Dave saw it, too and tried to lighten the air. He wasn't out to offend a single person through this exercise. He knew that there were huge problems with this process as it stood, but he also knew that he was the new guy and that this was a very proud team. He just wanted to teach them, to coach them, to make them think. He knew already they could easily fix whatever was wrong. He just needed to get them to see.

> "Hey, listen, I understand. I don't want anyone to feel wrong here. I know this was rushed. I can already tell that you're under a lot of stress and are under-resourced for what you have to do. It's also obvious that all of you really care. Some of the gaps here aren't that unusual. I can see a lot of great work has already gone into this. We just want to make it great for the operators and our customers, don't we?"

The tension fell precipitously, and he could sense that even George was loosening up a bit. Capitalizing on this, and realizing that the most important thing he could do, especially as the new guy, was to win hearts and minds, he said,

> "George, Pete has been walking me around the plant this morning. I saw some great things that you and the team have done, especially the new washer system. It's small, flexible, low-cost, and in line with the process."
> He turned to Pete, and said, "It really is fantastic." George had designed that one himself, and smiled broadly.

> Tension defused…for now.

> "Are we working on the standard work?"
> George responded enthusiastically. "Yes, I told you we'd have the work instructions for next week. I think we can pull them ahead."

Dave decided to let it go for now. He knew how critical standard work was. He'd not seen much evidence of it anywhere he'd been this morning. He was beginning to wonder what he'd gotten himself into. Then again, when he'd recruited him, Pete had been very open about what Dave would be facing. He suspected that things were worse than even he'd thought. But he'd seen many teams in the past make amazing improvements over time, and he was confident that this team would do the same. He knew he just had to coach them well.

"So, listen," he said, "I just want to put myself in the operator's place with this
air guitar exercise. Our mission is always to make the operator's work life a
lot cleaner, easier, safer, and a lot less frustrating. In fact, why don't we get
an operator to walk us through the exercise?"

Pete realized they didn't regularly get operators involved, but he was up for anything, and asked Mike, the shop manager, who was going to be working this station.

Mike said, "Johnny. He's our best assembler."

Our best assembler, yes, Pete thought. But he also tends to criticize things, which
might not be good. Before he could finish the thought and find a way to postpone
risking strong operator backlash, Dave said,

"Great, let's get Johnny over here."

Mike wandered off and came back a minute later to introduce Johnny Cox, a
seasoned assembler, to the team and to Dave, who looked him in the eye and
heartily shook his hand, welcoming him to the exercise.

"You know, all we want to do here is to just simulate the work. Let's just
observe, think about what we see or don't see, and maybe some things will
jump out at us that we can learn from and make this process even better for
guys like Johnny here. Hopefully, before we even start on Wednesday," and
he patted Johnny on the back. With what Dave was seeing, they could probably use a two-inch thick binder to write down everything they'd find.

But that wasn't what he wanted to do now. It was just the first of many long
days. He could already tell that it would be a long journey to coach his new
teammates just how to right their ship, let alone to get it to sail smoothly. Johnny
was grumbling already that there was no way they could possibly start up this
section of the line this afternoon. Nothing was ready. Pete's shoulders sagged.
This really was not going to be good.

Dave clapped his hands again enthusiastically, and said, "Okay, we have a lot
of air guitar to play. Let's get at it."
"That sure was a long walk over here," Dave laughed, feigning catching his
breath.

George was about to provide some reason for the distant storage area, but he
paused. He could see that everyone else got the message. Even he got it. They'd
been so rushed that they hadn't given it a lot of thought before, but this simple
exercise was making a lot of things clear. Nobody else was upset, but they didn't
say anything in response to Dave's quip either.

George and the two Lean leaders, John and Gina, were already beginning to think that they really hadn't thought this through well. They were starting to get a little worried and were feeling the defensiveness creeping back in.

Dave broke in again, "So how many different frames are supposed to be here?"

Mike, the shop manager, who'd been pretty quiet up until now, spoke up. "There are four types of frames. Two are common, and the other two we use a couple times a week."

"So where do we put each of them?"

"Right here."

"Right here?" Dave pointed at a broad expanse of barren gray concrete next to the main aisle.

"So do we separate them? Do we mark them so the material handlers know where to put them?"

Mike answered, "We didn't have time to think that through."

"Okay. Do we at least know how many of each we're supposed to have here?"

Blank stares. Eventually, someone spoke up. "We didn't have time to think about that either."

"That's okay. So how do you know which one to grab?"

George said, "For the next one coming down the line."

"How would I know that?"

"You'd have to look upstream on the line at the next unit's markings to determine what it is."

"Is it easy for me to see a difference?"

"Well," said George, "sometimes they all look the same, but an experienced operator can just walk back to the next station, look underneath, and tell which unit it is. And they know by experience which frame goes with which unit."

Even as he said it, George knew how inadequate it sounded.

The other two Lean leaders were cringing. They knew better than this, but they had overlooked putting in place a simple means to identify the frames that didn't require highly experienced operators. There just had not been enough time to think any of this through. Yet here they were, and in just ten minutes they had already exposed plenty of problems.

What Dave said was, "That's good." What he was thinking was, what a mess!

Pete, too, was silently cringing. He had been so confident in the six months he'd been there that they were making progress. And they had been, but he was starting to get the feeling that the problems were far worse than he'd thought. But the air guitar thing that Dave was walking them through was good, he thought, and he was comforted that his friend had joined them. It was only three hours into the first day and he had already learned a lot.

Dave didn't ask all his questions yet. Now wasn't the time to probe each weakness he saw. This morning would just be a first pass, to help the team realize that they have a lot to learn and a long way to go in the process—but also that this stuff isn't that hard.

"How heavy is one of these frames?"
Mike answered, "About 75 pounds."
"So, how am I supposed to get it over there?"
The team members looked at each other. Finally, Mike said, "Well, we have a similar frame assembly process in another line. There the operator asks another operator to help him carry it over."

Dave paused and looked at everyone. He didn't have to explain the safety issue that this represented, and not just on the new line. It was just one more big oversight that needed to be rectified.

"How many frame assemblies are we expecting to make each shift?" Dave asked.

Mike mentioned that they make about 40 units a shift.

Dave responded, "So the operator has to do this *40* times?" emphasizing the number 40.
"Johnny, give me a hand carrying this frame over there."

Dave and Johnny both bent down, simulating lifting the imaginary frame, and started walking sideways over to the station and up to the platform, again exaggerating the exertion the simulated carry required.

"I guess we carry this up there?" Heads nodded all the way around. Dave grunted. "Let's heave it up there, Johnny."

Dave and Johnny put the simulated frame down on top of the fixture that would be used during the assembly process and was actually in place on the transport cart. As they were lowering it, he could see that the cart/fixture combination had been well thought out. It was creatively simple, very flexible, and low cost, yet effective. He smiled, thinking that maybe there was some promise for this place after all, and praised George and his Lean team for the design.

"Johnny, get up here beside me; it's time for a lesson."
Johnny quipped, "You're going to teach me how to assemble?"
Dave responded, "Nope. You're going to teach all of us how best to do this. But first, I want all of you to think of the operator as a surgeon."

OPERATOR AS A SURGEON

Johnny is like a surgeon in a hospital. Every other person in the hospital, particularly the surgical nurses, is aligned around supporting the surgeon's efforts. Imagine that you were a patient on an operating table, and the surgeon had just made an incision. Now imagine that your surgeon regularly left you alone on the table to find and retrieve medical supplies, tools, and equipment. How good would you feel if this happened to you? As a patient, the two biggest objectives you have are to get off the table as fast as you can, and not to have to go back onto the table for any reason. The same holds true for operators on a line. The operators are the surgeons, and the work-in-process is the patient. The rest of the organization should be aligned around supporting the operators so that they can add value. They should not have to leave their work area.

Dave asked, "So what do you think about that?"

Mike chimed in. "Well, it makes perfect sense. Our operators spend a lot of time just looking for parts. The downside is that we would have to hire a lot of new people to do this, and we can't afford it."

Out of the corner of his eye, Dave saw George nodding in agreement. The other two Lean leaders were shaking their heads as though they had dealt with this before to little effect.

Dave said thoughtfully, "I can see how you'd think that, Mike. But let's think about a few things. How many people do you have on this section of the line?"

"Forty per shift."

"And how much time do you think they spend getting and looking for parts?"

"I'm not sure of the exact number."

"That's okay; we don't need an exact number. Just a ballpark is fine."

"I'd say 20%."

"Good. Who knows what's exactly right, but as we've been standing here and looking downstream at the working line, it seems to me that at least 10% is a reasonable number."

Mike responded, "Seems reasonable. You're probably right."

"Okay. So if we use that number, 10% of 40 people is the equivalent of 4 operators."

"I never thought about it that way," Mike said.

"That's okay. Most people don't. Just think, if we got four folks off the line to get the parts for their peers and rebalanced the work on the line, we'd be no worse off. Plus, if we were to develop standard work for material delivery and simplify and improve the storage to reduce searching, we could maybe support this section with only one person, maybe two."

Dave was hesitant to give them too many answers. His experience taught him that it was more effective to share ideas to get people thinking without directly telling them what to do.

"Also think about something else. I've only been in the plant a few hours, but I see a lot of material handlers actually moving material one type at a time, with forklifts from point A to point B. How many material handlers do we have?"

"About 46."

"Whew!" Dave let slip. "Let me give you another thought."

TAXI VERSUS BUS ROUTES

How do people move around in cities? You can probably think of a number of ways. Let's look at taxicabs. A dispatcher might get a random, unanticipated call to pick up one or several people from a certain place and drop them at another place, and this happens thousands of times a day. Cities also tend to have buses with formally defined routes, multiple defined pick-up and drop-off points along the route, a schedule for the route that defines when they will be at each pick-up point, and they move many people at the same time. Most companies, like us, move material through their factories according to the taxicab model. I wonder what it would be like if we were to have defined routes, a standard work sequence of stops with timing, and moved many different types of material around the route, including the stockroom, using a tugger and a train of carts behind? Does anyone think that would be better in many cases? Do you think it would be possible to try something like that here?

"I think we'd have more than enough material handlers. What do you think?"

Dave saw all heads nodding, even George's. The two Lean leaders, John and Gina, looked animated.

"I love this stuff," Pete said. "This is just great."

"So let's get back to air guitar. Stay up here with me, Johnny."

"What do I do with this frame now?"

"You attach a cowling with four bolts."

"How do I do that?"

"With the air driver," George responded.

"That one there on the floor with the hose lying around? Does that look safe to anyone?" Silence.

"Where do I get the cowling?" George pointed over to the aisle, closer than the frames.

No area was marked out, and the operator still had to come down off the platform and get it, which Dave did, bringing back the simulated cowling after asking about type again. The same issues they'd found with the frame were found with the cowling.

When Dave got back on the platform, he asked,

"What bolts do I use?"
"1/4-20," George responded.
"How do I know?"
"We'll train you."
"So I'm supposed to remember…. Oh. There are work instructions, right?"
"Right."
"So where do I get the bolts?"

George pointed to the bin at the station.

Dave smiled and said, "Good," and grabbed four bolts. He asked how these were replenished. The grimace from Gina told him that this was one more thing that had been overlooked.

Dave reached down and picked up the air driver. "Wow, this is heavy. How long does the operator have to do this?"
Johnny said, "We have to use it all day. You can get real sore by the end of your shift."
"Mike, why don't you try lifting this?"
"Wow, this *is* heavy. I had no idea."
"That's okay, Mike. This new journey will be all about learning, and all about making Johnny's life easier."

Johnny smiled; he obviously liked the sound of that. Lean leader John Lee offered that they could probably put it on a balancer hanging off the pipe above.

"Great idea," said Dave. "Now you're talking." He was pleased with this spontaneous and effective solution.
Pete walked up to Dave. "This is going great, but we have to meet our inside sales manager at 11:00."
"Okay," said Dave, "we didn't get through anywhere near all of this today, but what have you learned so far?"
John Lee said, "Your air guitar process is easy. I can see how you could do this on any process at any stage of development, even if we haven't got anything in place yet."
"Great observation. You're exactly right. This is a key element in the full Toyota 3P process. We'll be introducing the whole 3P concept here one day. Sooner rather than later, I hope. What else?"
Gina Nelson spoke up. "We still have a lot of work to do to get this line expansion into a strong process for our overall operation."

"That's another great observation, Gina. It sure seems like we do. But listen, everyone; Lean is about trying things and learning as you go, getting smarter and better with time, and continuously improving. I'm sure we all would have wanted more of this in place at this stage, but you recognized a problem with capacity and started this whole new system on your own. That's a great start."

Patting George on the shoulder, he continued, "There are some really good thoughts in the current design; I can see that, even though we didn't review all of it yet. But of course we can do better, and next time we will."

"Why don't you guys go through the rest of the processes the same way? You don't need me," Dave said. "Even though this is just a simulation, make sure you ask, do, and observe down to the smallest detail, and keep lots of notes. We should try to get as many of the easiest ideas done before we get too far into the pilot this afternoon. Our Lean journey will be about moving fast with improvements. In fact, if I were you, I would think about getting some maintenance folks over here now, while you continue to simulate, and get them working on some of the gaps we've already uncovered."

George said slowly, "We can do that."

George sighed. He had to admit, most of the discoveries had been good ones, and he liked the new "air guitar" method of simulating the process. Still, he was defensive and embarrassed about the development process he'd led. But the new guy seemed to know his stuff, and it didn't seem like he was trying to make anybody look bad.

Both of the Lean leaders were thinking almost the same thing, but with no small measure of relief. Maybe they hadn't been quite as skeptical as George, but they still hadn't greeted the news of Dave's arrival with unequivocal joy. Sure, they were strapped for resources, but hadn't they been doing okay on their own? What would this new boss—a Lean expert, no less—think of them? Would he change everything? Were their jobs safe?

But, already, on Day 1, their anxieties had been almost entirely allayed. They could see that Dave really got it, that he was good, and that his style would really work here. Maybe, just maybe, they could finally get the naysayers and fence-sitters to start contributing more positively to moving things forward. As he was walking away with Pete, Dave patted George on the shoulder once more and said,

"George, you're in charge. Go to it!"

"Well," Pete said, "let's go take a look at the transactional side of our operation," and off they went. "I really like the air guitar simulation exercise. We can use that on a lot of new layouts. What did you call it? 3P?"

Dave responded, "It's part of 3P thinking and it can be very effective, but it is not full-blown 3P, which is even more powerful." Dave continued, "I hope I can get us into that in a big way one day. We don't seem to be good at introducing major change."

"That's the truth," Pete said.

"By the way, we can do air guitar in a lot more places than just the factory. We can simulate information flow processes, or material or production flow. I have a friend who does this all the time in the hospital that he works at."

"Really? That's impressive," said Pete.

"It sure is," Dave replied. "I'd like to see us do it on order entry."

"But, that's an existing process, not a new station," Pete said.

"That's okay. We can apply air guitar as effectively and easily on an existing process as on a new one."

"Really?"

"Really!"

Pete said, "Then let's see how our inside sales processes hold up to air guitar."

"This is Mary Long, our inside sales manager, and Sylvia Bennett, who oversees the order entry process and reports to Mary," Pete said. "Ladies, this is Dave, our new Lean leader."

"Nice to meet you both," said Dave.

"I'd planned to have Mary describe our order entry process for you," Pete began.

Dave cut in. "I'd rather just follow the flow of information from the time we get an order from a customer until the time we have it entered completely into the system."

Dave liked Mary immediately. Almost apologetically, Mary said,

"We already processed all of our orders for the day this morning. Unfortunately, it was a pretty light day."

"If you just walk us through all of the normal steps, we can simulate how the order information flows. That will give me a pretty good understanding of how things work today. I call it 'playing air guitar.'"

Mary and Sylvia laughed and walked them over to a set of cubicles. Although the total area was fairly small, Dave thought that it looked like a maze and wondered how anyone could easily communicate with each other. Mary pointed out that there were four associates who first process the orders. They each had a cubicle, each isolated from the others; they may as well have been in a different room.

"Where is everyone?" Dave asked.

Mary responded, "As I mentioned, today we finished early. Most of our orders come in in the morning, and today we had fewer than normal. Sally is here to handle orders that come in the rest of the day. The other three associates keep themselves busy either helping other associates or handling rework. We get a lot of rework; our salespeople and customers just don't seem to enter the orders correctly. It's complicated, but we'll do whatever it takes to get it right. That's why they like us so much," she said ruefully.

Dave said, "That's great. Focusing on customers is one of the most important things a person or company can do. It'll be one of the key principles of our Lean journey here. Unfortunately, not everyone does, because they tend to be internally focused."

"We do," Mary and Sylvia chimed in together, proudly.

"Maybe with time we can all figure out how to simplify things for our sales team and reduce some of that rework. Right?"

Dave turned to Pete, who was watching attentively.

"Right," he replied.

"I'd love for us to do that if you think we can," said Mary.

"I'm confident we can," Dave said with gusto.

"So if I'm an order from a customer, how do I get here?"

Sylvia hesitated. "Well, there are four ways, really. We can receive them on the fax machine here. We can also get them in the mail. Most of the time the mailroom puts them in that tray over there, but sometimes they divide them among our desks. Some orders come over the phone; we take turns answering the phone during the day. And customers can send an order for some of our products directly into our system via the Internet. We still inspect them when they come in this way to make sure the customer entered the correct information and that nothing's out of the ordinary. We're now getting about 30% of our orders this way. Half are still mailed in."

"Thank you Sylvia. That was helpful," said Dave.

"Let's pretend I'm an order coming in the mail. Let's follow the process. Does anyone have a blank sheet of paper?"

"Here's one," Sylvia said, curious about what Dave was doing.

"So I assume this comes up from the mailroom."

"That's right, usually around 10:35."

"Let's pretend that I'm the deliverer and I'm walking with this order."

Dave walked over. "Okay, what do I do with it?"

"Well," Mary says, "it usually comes over with 10 to 30 other orders, and most of the time they put them in the tray here. Sometimes Rick will sort them by region and set them on each of our desks. John never does; he just throws them anywhere."

Dave recognized the lack of standardization, but said nothing.

"If they're always in a different place, how do we know how we're doing on processing them, and don't they ever get lost?"

"Yes, all the time, but we almost always find them. For the processing, we just all work through them until they're done. Come to think of it, the guys in the shop are always complaining that when they finally get the order released to them, it's overdue. Pete brought a consultant in here a few months ago for a week to help make some changes. Remember, Pete?"

"Yes," Pete said, not happy at the reminder of that person and the outcome. Mary went on, "But we still have many of the same issues."

"Besides," said Mary, "we've been taught that the customer needs to be our biggest focus, and we work hard to look after them."

"I can tell. One of the things most customers look for today is a supplier who is fast and reliable on their deliveries. One of the things we want to do with the Lean program that Pete has tasked us with is to move our information through a lot faster and smoother."

"Pete told us," said Mary, "and we're really excited by that. We're the ones who receive the complaints from the customers when they're unhappy."

"And we don't like receiving those phone calls," added Sylvia.

"Excellent. Let's talk about the customer a little more. Who are your customers, Sylvia?"

"Oh, that's easy. I handle everyone in the northeast."

"What about the shop floor folks?" Dave asked.

"What about them?"

"What would they say about our order entry process?"

Sylvia responded, "They complain a bit, but mostly we get along well."

Time for a teaching moment, thought Dave.

INTERNAL CUSTOMER

A lot of times, people consider the person who ultimately buys their product or service as their only customer. Although the external customer is obviously the most important customer, it is not the only one. If all of our employees thought of the next person who they pass their work on to as their customer, and treated them as such, our performance would skyrocket. Internal customers want the same things that external customers do—good quality work, when they need it, and to be treated decently. I often challenge people to think of themselves as being self-employed at the work they are currently doing. Then imagine that internal customers could choose from whom they get their inputs. Would your business be thriving, or would your customers go elsewhere?

Dave moved on. "So which of these trays is the inbox? I see four here."

"The one on the left. John can never seem to get them right," Mary sighed.

"Why don't we label them now, so it will always be easier? We might call such changes 'Just Do Its.' Too many times we're too busy, and simple improvements just don't get done. If a team gets in the habit of just doing the simple and obvious improvements, over time it can make a big difference." Sylvia nodded and went off to get a label maker.

Mary jumped in, "You know, we probably should insist that the mail deliverers always put them in one of these bins, so we know where they are and can track them more easily."

"Now you're talking," Dave smiled broadly. Pete smiled too.

"Okay, so if I put this one in the tray, what happens next?"

Sylvia offered, "They get picked up and put on the right desk whenever someone comes by and sees orders there."

"So at any time they can get picked up?"

"That's right," said Mary.

"So, where would this one go if it was from the northeast?"

"It would go on Sylvia's desk."

"How would I tell it was from the northeast?"

"Well, that's where it's difficult sometimes; the orders are random in terms of customer format, the address can be anywhere, and we have to decide if the shipping and billing addresses are different and which one applies. Different people decide differently, and many times wrongly. It's also not easy to quickly tell which region the address is in, and often we have to take the time to get a map and see what city and state are assigned to each order entry person."

"Sounds like it could take a while to do."

"It can. It's a pain to sort through them some days."

"Okay, so where do I put it on Sylvia's desk?"

"Just find a spot," Mary said.

Dave lowered his eyes and suggested that maybe they could add a tray as an inbox.

Sylvia said, "That's a good idea. It'll keep my desk more organized. I'll do it right away."

"So now what happens?"

Sylvia said, "I work through the faxes, electronic orders, and phone calls as they come in, and I squeeze these in when I can. When I get to them, I enter them into the computer. It usually takes between 20 and 30 minutes to enter an order."

"How long can the order sit here?"

"Sometimes I do them right away. Usually, I get to most of them before the end of the day, but not always. I try to get to the backlog by the next day."

Dave nodded his understanding. "Wouldn't it be nice if we could figure out how to smooth out the flow?"

"It sure would," Sylvia said. "I'm always stressed out, even on slow days. I'm always rushed, and I know I make mistakes. Do you really think you can help with this?" Sylvia asked.

"I'm 100% confident that with time I can help the two of you figure this out on your own. In fact, over the next few weeks, why don't we learn a few new tricks?"

"Sounds good," Mary said. "We'll be looking forward to it."

"So now that the order is in the system, what happens to it?"

"We have four sales reps who go in and pull up the order on the screen and check through all of it, especially for price, credit, and configuration issues."

"Where do they do this?"

"They sit over by the cafeteria on the other side of the office."

"Okay, let's go over there and see."

"So, they up pull up the order and inspect it to verify that it's correct. What if they find a problem?"

"They reject it and usually call us to talk about it, if they can get us, but we can't always pick up."

"How many get rejected?"

"About 25% of them."

Pete looked stunned. "Wow, I had no idea."

"So, what if I pass through?"

"Oh, that's easy. It goes back in a queue, and our engineering techs go through and do any necessary adjustments."

"How long does that take?"

"About 30 minutes an order."

"Then what?"

"Well, if they don't find any problems, it's ready for the master scheduler."

"This has been very helpful; thank you. Why don't we all go grab some lunch, and then we can set up a little simulation to help make things better."

After lunch, they gathered around the order entry area at an open table.

"So Mary, let's get four trays and put them across the table here. Let's label them with masking tape: 'mail order,' 'electronic order,' 'fax order,' and 'phone order.' For now, although I think it would be better to just flow them all together, let's assume we have a process that presorts them into these trays. Let's start with only northeast orders to make it simple. Let's get three folding tables in a row here behind these four trays. Mary, why don't you sit at the first table and do order entry. Sylvia, you sit at the second table and be the order checker. Pete will be at the last table to do the engineering checks. By the way, if we designed and configured our product more easily, we wouldn't need you here, but maybe one day! Let's put two trays at the front of each table and mark one 'Input' and the other 'Rejects.' Okay, how many northeast orders do we typically do in a week?"

"It can vary, but between 75 and 85."

"Okay, let's change the label on the Input tray at Table 1 and call it 'Accumulated Input Backlog.' Let's put a line inside here at about 10 sheets high and another mark at about 30. I'll explain that in a minute. If we do an order every 30 minutes, how many could we do in a week?"

"Eighty," Mary said.

"So why don't we try and do one every 30 minutes?"

Sylvia offered, "I think there are times that we can do our part in less than 30 minutes. Some may take only 20 minutes."

"That's okay. Let's just consider that a stress relief for now."

"Okay. I'm going to take 20 pieces of paper and put them in this 'Accumulated Input Backlog' bin. If we get fewer than 10 because orders come in more slowly than the average of 30 minutes, we'll stop; and if we get higher than 30 because we're slower than 30 minutes too often, we'll also stop. Okay?"

"Okay," said Mary, Sylvia, and Pete.

"So, let's load one order in each inbound tray. I have a stopwatch here and I'll time it. We'll use 30 seconds as representing 30 minutes. I'll call go. Each person will write his name, address, and favorite food on the sheet and then raise his hand. Write neatly and don't rush. I'll call move time and if everyone's hand is up, you'll each pass a sheet to the input tray on the next table. If someone's hand isn't up, we'll wait until he's ready until we flow the sheets down the tables. Then we'll start the 30 seconds all over. Everyone ready? Go!"

After ten passes through—only three minutes of simulation—everything was going smoothly. Dave hadn't expected anything different. He didn't introduce rejects. His intent was just to get people thinking about new possibilities—a few significant but simple changes, like breaking into teams by region instead of by function.

"So what do you think?"

"Well, it was easy, but that's not the real world," said Mary.

"Tell me more."

"We don't sit beside each other and we don't work together by region. Many times, we get rejects at each station, and sometimes the orders take more than 30 minutes, and they don't come in to us at a consistent rate. Sometimes we'll get a phone call, and we have to do the order right then," Mary said, without taking a breath.

Dave laughed, "But it sure was less stressful, wasn't it?"

Sylvia laughed too. "It sure was!"

"All of these things, along with other things I've observed, are real barriers to easy flow. Pete, if we have another hour or so, I have a lot more simple tricks we can play with. We've only been simulating for a few minutes, and you guys have had these struggles for years."

Pete said, "Well, we'd planned to tour you through engineering. What did you have in mind?"

"If we keep experimenting with this simple simulation, we could all work together and solve all of the issues Mary brought up."

And so for the next two hours they shared ideas, trying some of them, and starting again. In the end, they had a simple process that seemed to work well for any issue they could raise. All orders went through the final simulation. They determined that they could easily complete 80 per week, with one person handling the phone calls, which would allow the other two to work more smoothly with fewer interruptions. All three could work on the rejected orders when time allowed, such as when the rate of orders coming in was low. The real key was having all three people physically together working as a team for each region.

Mary and Sylvia weren't sure that it would work in practice, but they all liked the proposed process and felt there were some obvious improvements that would make things easier.

Dave said, "Why don't we try it?"

Mary asked, "When?"

"How about this week? We could create an area and try it just for the northeast. I'll bet we could get set up in less than two hours. What do we have to lose?"

"Nothing I can think of," said Mary, "but the guys in the other areas probably won't want to move."

Pete said, "Well, they're just going to have to try it. They'll get used to it, and I'm sure it'll make their work life easier as well. Let's do it!"

Over a few drinks later that evening, Pete said, "Dave, I can tell you, I'm so happy you've joined us. In just one day, I can see how much you can help us."

"Pete, I'm excited to be here and to be working with you again. You know, there's nothing that I saw today that we can't fix together. You've got a great team, and with time we'll bring them all along—even George, who's still a little skeptical."

"To say the least," Pete murmured.

"You know, Dave, this air guitar stuff can really help us, and we can do it on any of our processes."

"I think so, too," Dave responded.

And so Dave, Pete, and the rest of the team began their journey to fix the Memphis operation.

Meanwhile, the CFO was working with his team to develop a new strategy to convince the boss to relocate it to Mexico.

Chapter 3

The Moonshiner

Romano's was the hottest new Italian restaurant in town. Located in the heart of the entertainment district, its lines typically started to form at 5:00 p.m., seven days a week, with crowds hanging around its outside pillars and fountain. Sitting in a corner booth were Dave and his wife Julie. He'd been working a lot of hours since starting at Trail Rider, spending weekends with the kids, and taking quality time with his wife whenever he could.

"So how are you *really* finding it?" asked Julie. "Do you think that you made the right decision? You've seemed a lot happier the past few months."

Dave responded, "I love it. We're making a lot of progress, and Pete has given me free rein to make changes. We also have this new sensei who's been teaching me a lot."

"I thought you were the Lean expert," his wife jibed.

Dave said, "Nope, I'm still a student. In fact, I'm finding myself surprised at how little I know." Changing the subject to their daughter, Dave said, "Tess seems to be handling our move to Memphis well."

Julie responded, "She's fitting in really well at the new school. Struggling a bit with this bossy girl in class, but she'll handle herself."

"Just like her mother," Dave said.

❋　❋　❋

Debbie Grant brought Pete some barbecue sauce and a fresh beer out to the grill, where he was just turning over the thick-cut New York strip steaks he'd picked up in the afternoon.

"I don't think I've seen you so relaxed in months," she said. "Things must be going better at work. Dave seems to be really helping out."

"They couldn't be better. I was worried after I started. Things were bad and just kept getting worse, and nothing I did seemed to work. Remember the pressure I was feeling from corporate? Well, things have been quiet the last month or so. I think that's a good thing.

"Dave is the best decision I've made," he continued. "He's working out really well. How about you? You've been traveling a lot with work the last few weeks."

"Yes, but that'll ease up the next few months. The accounting biz isn't what it used to be. You'll be seeing more of my smiling face around here. In fact, let's go out with Julie and Dave next weekend," said Debbie. "Mention it to Dave at work when you see him tomorrow."

"Will do," Pete said.

❋ ❋ ❋

Three months had passed since his boss had given Trail Rider until the end of the year to turn itself around. Steve Sawyer, the vice chair, sat at his desk going over September's results and broke into a smile. For the first time in three years, his troubled business had turned a positive profit for the quarter; better yet, their operating margin was up 25% over the prior year. He made a note to call Pete Grant. Just then, he saw the CFO walking past his door, and he yelled out,

"Hey, John, did you get a look at the September financials?"

John knew exactly what Steve was getting at. He grumbled,

"Anyone can make a quarter. They didn't exactly have a stellar quarter last year so it didn't take much to get better." He couldn't resist a parting shot.
"If they were in Mexico, imagine how much money we would have made!"
Steve said, "It would be nice if you could give our employees credit where credit is due." If he'd hit a nerve, John did not acknowledge it. He said nothing and continued his walk down the hall.

❋ ❋ ❋

Around the same time Steve Sawyer and John Cuso were exchanging words, Dave Martin was on one of the engine feeder lines working through a line rebalancing issue. They were expecting a change in volume and product mix over the next two months, and the line needed to be adjusted to meet the new rate of production, or "takt time" as Lean practitioners called it. He reflected on the progress he'd seen in the plant since he started. The simulation exercises had been spreading, and people were stepping up, trying them out in their own areas and coming up with ideas for improvement. In fact, he was out on the line now to show the Engine Team how they could quickly check their proposed line-balancing scenario by running a simple simulation using small tiles that represented production units. They had nine stations on their line and two different engine models: Model A and Model B units. They had about the same work content at all stations, except Station #6, where the process time was 20% more on Model B units. This had't been a problem at the rate of takt time they were currently running, but they needed to increase the production rate by 10. This caused a problem for Model B units when they arrived at Station #6. They needed to develop a new approach and new standard work to accommodate the increased percentage of the more complex Model B units they were to produce, and the higher rate of production they'd be running for at least the next 60 days.

John Lee, one of his Lean leaders, had suggested that because all the other stations were fine, they could add another station and another unit of work-in-process (WIP) to the line at Station #6, creating a Station #6A and a separate Station #6B, and expanding the line to ten stations. Then, when a unit of the easier-to-build Model A engine flowed down the line and reached Station #6A, all work on it would be completed at this station. When that unit was completed, it would then move to the new Station #6B and sit in a queue before it moved on to Station #7. When a Model B unit, which required more time to complete, was being processed, it would have the majority of work performed at Station #6A and the rest at Station #6B. This would provide the extra capacity needed to process the more complex Model B units at the desired rate of production, or takt time. The only issue Dave saw was that the team would have to learn to flex a person in a "floater" role to Station #6B when a Model B unit was there, and move them away when a Model A unit was there. The team was split and began a spirited debate about whether or not it would work.

"Let's try it," said Dave. He cleared off a table along the line and laid out ten Post-It® notes, representing ten stations, quickly identifying each with a number. He asked, "What's the percentage of Model B units we need to produce?"
"Thirty-three percent," someone shouted out.

Dave grabbed 15 small square tiles—10 white tiles to represent "A" models and 5 red tiles to represent "B" models. He placed them in the simulated line; two As in a row, then a B, all the way down the line, one tile at each station. Another five tiles were placed before the simulated line in the same sequence. These represented incoming work. Dave explained to the team that, every ten seconds they would advance the tiles one station ahead. As one tile exited the line at the end, a new one would be added at the start. This simple simulation would allow the hourly employees and the supervisor to see what happened when a unit of Model A and a unit of Model B went through Stations #6A and #6B. They were amazed at how simply and fast this exercise allowed them to test the proposed idea. They quickly saw that it would be quite easy to flex someone into Station #6B to keep the flow going at the desired rate. The grumbling died down. The team seemed satisfied, and John Lee mouthed a thank you to Dave. As they went back to work, Dave recommended that the team keep the tiles handy for future simulations.

Dave lingered with John Lee, as he hadn't spent as much time in this area as he had in others. Together they watched for about 30 minutes as the line started up. Dave had noticed all kinds of little things that were problematic for the operators. He saw people struggling to lift material to the line, an operator bent over at an awkward angle with a wrench trying to turn a bolt from underneath, and another operator reaching through a shelving unit to get a ten-pound part from behind it. He saw the opportunity he was looking for and asked John to come with him to find Pete Grant.

Pete was in the cafeteria, leaning against the Coke™ machine and talking to one of the hourly team leaders.

Pete looked up and asked, "What's up, Dave?"

"Do you have a minute, Pete? We'd like to talk to you about something."

"I'm just wrapping up here. I'll meet you in your office in a few minutes."

A few minutes later, Pete wandered into Dave's office, where he and John were waiting.

"So what do you want to talk about?" asked Pete.

He could see Dave's brain churning, and he knew he was in for yet another coaching session. For the first time in years, they were turning things around. But he had to admit, he often felt like a sponge being hit with a garden hose. Dave was a dynamo and kept everyone hopping. And although they'd made improvements, it had only been three months, and things had been running very poorly beforehand. It didn't take a lot of improvement to make a real difference. They still had a long way to go just to get out of the woods and reach even a semblance of average, let alone good, performance. Still, the momentum was swinging their way, and he was pleased.

"Moonshining. Ever heard of it?" said Dave.

Not a clue, thought Pete. "Sure. That's the stuff they distill in the mountains, isn't it?"

"Not even close," said Dave. "Moonshining is the practice of developing simple, low-cost tool and equipment solutions to problems the operators are having, or wastes that exist in the process. Mechanically and electrically oriented hourly employees, usually out of maintenance or the tool department, will build things from scratch, without drawings, using a lot of used materials, parts, and assemblies they find lying around. In the beginning, they often design and build simple solutions to make things safer and easier for the operator. As they get more advanced, they build a lot of the advanced tools, fixtures, transport devices, furniture, and equipment. In fact, some plants call these folks 'equipment design build teams.'"

Dave paused for breath. "More mature companies design and build a lot of their own equipment internally, often saving 30% to 50% on plant and equipment capital expenses over what they would typically spend. They can do so in much less time, and, as you know, time is money."

Pete listened intently as Dave went on.

"They design equipment that is right-sized, geared specifically around the application it's intended for, has few to no bells and whistles, and is easy to service and maintain. The tools and equipment are often much better received by the operators because they were built by a colleague who listened to their specific needs. In most plants, manufacturing engineers, who are smart and always well-intentioned, often reach outside to off-the-shelf or custom-engineered solutions that typically cost a lot more and are much larger and more complex than they need to be. Off-the-shelf solutions don't always meet their specific needs, and custom-engineered solutions take a long time to acquire. When a manufacturing engineer introduces an off-the-shelf new piece of equipment, how do you think they're received by the operators who have to use them? Eye rolls, usually. Pete, I think we could save half a million or more per moonshiner if we put them in. Plus we'll significantly reduce turnaround time."

Ken Saguchi poked his head around the corner and said, "Hi, everyone." Ken had come in a few weeks after Dave started, for a week each month, under a consulting contract arranged by Pete. In the past three months, the three of them had gotten along famously. He'd helped Pete and Dave develop an improvement strategy, had facilitated an assessment and process re-design exercise using value stream mapping, and was leading the team through a series of kaizen activities. The three of them, with a cooperative workforce behind them, had started to get some traction on safety, quality, delivery, and cost improvements. They had a long way to go, but the early momentum had been well received, especially because they'd been struggling for so long.

"Just wanted you to know I'm heading out. See you next month."

Ken was thinking about heading home for the weekend. Working with Trail Rider had been the tonic he needed. Seeing the Memphis operation engaged so passionately in a mission to improve had raised his spirits, which in turn helped his relationship with his wife and family.

"Hey, Ken," Dave shaking him out of his fog, "tell Pete what you told me about the ideal moonshiner. I've been talking to him about how we might use that idea to help drive some big improvements in many areas."

"Sure." said Ken, "Remember the 3P concept that we've been talking about using to implement major product and/or process changes with fewer issues? Well, strong moonshining will be a real help with any 3P effort. If you start strong and grow capability, it'll pay off in many ways in the future, especially if you embrace 3P. Let me describe the ideal choice for a moonshiner."

THE IDEAL MOONSHINER

The ideal moonshiner is typically a very mechanically oriented hourly person, creative, thinks well on his own, listens to others, and understands. You're looking for the kind of person who lived with his family on the farm and worked with his dad on weekends tearing the transmission out of the tractor, troubleshooting and fixing the hay baler. They fixed a lot of the problems around the farm with limited resources, using whatever they had lying around.

Pete said, "We have a few employees in maintenance who are just like that. Bill Cook is a toolmaker, and he's built a lot of pretty ingenious fixtures for us over the years. And there's Norm Wilson in our maintenance department. I sat with Norm at the last Christmas party. He's been rebuilding cars for the past 20 years, has a big machine shop in his garage. His wife said he's always tinkering with things out there and has built a lot of equipment for their church group."

Dave asked, "Do you think I could tap them for some of this work?"

Pete responded, "Sure. Let me run it by Harry in maintenance. I'm sure it will be okay. In fact, I'll make it okay. I like the possibilities. Ken, you're getting me more and more intrigued with this 3P stuff."

Ken said, "I'll see you guys when I get back. Good luck with the kaizen we planned."

"Enjoy your time with the family," Pete responded.

"Thanks, Pete."

A few hours later, Dave and John were back on the line with two of their newly anointed moonshiners, Bill and Norm, neither of whom knew or understood what was being asked of them. But Harry told them to go with this Dave guy and cooperate with him, and Harry was the boss.

Dave was excited to get started.

"Listen, guys, Pete told me that you're some of the best mechanical minds in the company. If you spend some time with me in the processes, you'll see a lot of our colleagues forced to do work that is hard, tiring, and risky to their bodies and health. And many of the tasks they have to do use equipment, tools, and fixtures that are not ideal for the application, take longer to use and are more costly as a result, and can cause quality issues."

Now that their curiosity had been piqued, Dave went on.

"We want you to observe opportunities to make work safer, easier, faster, and
better for the people on our lines. We want you to work with our operators
and talk to them about how you can help them with their work. Take any
ideas you have to improve things, as creatively as you can, go back to your
tool room, and play around with solutions. Spend as little money as possible.
We know you can grab a lot of stuff around the site, from scrap metal to old
motors, pumps, and switches. Don't be afraid to make mistakes, and take
a risk on some solutions that may not work at first. There's no such thing
as failure in this game. The key word is 'trystorming.' If something doesn't
work, learn from it and try again. Failure is just a bridge to better ideas. It's
also important to work with the operators as you develop solutions. They
can suggest a lot of simple tweaks and improvements to anything that you
come up with. The idea is to start off with small, simple solutions, modifications to tools and fixtures, tweaks to equipment, etc. Over time, as you get
more confident and we get more courageous, we can take on bigger solutions. I'm not sure how far we'll take this, but I like to think there will be
a day when we build all of our equipment ourselves, rather than buy the
wrong stuff off the shelf, or get big outside engineering firms to design and
build it for us. Who could know better what we need than us? Who would
know better than you two guys? Any questions, thoughts, or concerns?"

"So," Norm said, "we fit this in with our other maintenance and tool and die
work? How long should we spend on this?"

"Great question," Dave said. "I've cleared it with Pete and your boss, and as long
as you enjoy this work, we want you to do it full time, every day. Others will
fill in on your current work. I think you'll love it, but please don't feel obligated.
I know you'll be good at it, but if you'd prefer, we can leave you where you are."

Bill asked, "Where do we do this work?"

"Another great question," Dave replied. "For now, just use the tool and die
shop, but we'd like to set you up in your own moonshining shop. Any ideas
on where we could do that?"

"Well," said Bill, thinking a second, "there's a perfect little room just off the
main assembly line. There's nothing but old files in there. If we cleared it
out, could we use it?"

"That'd be perfect," Norm agreed. Dave could see that they were both starting
to get into the role.

He said, "Let me check with Pete, but I can't imagine it will be a problem.
How big is it? You'll want enough floor space for the equipment you need
and storage area for supplies and old raw material, components, and parts.
Give me a list of what you think you might need. You can add onto it as
you get more into this."

Norm said, "There are a lot of spare pieces of equipment that we rarely use.
I can get a welder out of the old maintenance shop."

Bill said, "There's a mill machine across the road in the old warehouse. We don't need to buy a lot of new stuff—maybe just some tools for the equipment. You did say you wanted us to work low cost, didn't you?"

"That I did," Dave said, pleased as punch.

"How do we come up with projects to work on and keep ourselves busy?"

"I'm pretty sure that before long, being busy won't be one of your bigger issues. For now, the three of us will walk the processes and observe opportunities to modify or build things to make work easier. We'll let everyone know what we're up to and what we can do for them, and ask them to submit idea requests. We probably won't get a lot at first, but once you guys do a few things to make work safer or easier, word will get around, and more and more people will want your help."

Bill added, "We've run into problems before with a few things we did that helped put people out of work. I'm not sure I really want to go do this."

"That's a fair concern, Bill, and I don't blame you. Who would want to do things that put their colleagues out of work? I know you're probably skeptical, and maybe for good reason. I realize that you aren't just going to believe everything I say, but I'm not going to mislead you."

"We've heard that before," said Norm.

"I can appreciate that. And like I said, based on your past experience, you really have no reason to trust me. I'm going to look you in the eye and assure you that we won't let anyone go based on the work you do. If you ever think that's happening, you can go right back to your old jobs. This is about making work safer and easier for people, making them feel better about their work and the company, and hopefully helping them feel like getting more involved in making further improvements. It's also about driving productivity. But we're growing, we have a lot of temps here, we had a lot of overtime last year, and we can free up a lot of people and not have anyone lose their jobs, and still wind up with improved productivity."

Bill responded, "I get the temp and overtime issues, but I'm not sure about how the growth helps."

"Let's say we double our revenue. In ordinary times we would also double our workforce. Let's say we could find a way to still add people, but maybe not at the same rate as the growth. If we only added half the people we need, because we were more productive elsewhere, that's good for everyone, and it helps improve our competitiveness. Keep in mind too, that some people may change the type of work they currently do. They may get cross-trained in different jobs. That will allow them to rotate jobs for safety and health reasons, and to mix up the work to reduce the boredom and make it less repetitive. But we're not doing this so they lose their jobs here. Please trust me on this."

Bill and Norm nodded in agreement, although Dave suspected that they were still a little skeptical. He could also see that they were intrigued by the new opportunity.

"What else, guys?"

"Do we have to fill out timesheets of the work we do?"

"Tell you what. I want to keep it simple, and I trust you to stay busy. Just punch in in the mornings and punch out when your shift ends."

"What about overtime?" Norm asked.

"Well, I'd like to keep that to a minimum, but if we need it to support the teams on the lines, let me know and we'll take each situation as it comes up. Anything else?"

Bill replied, "When do we get started?"

"Right now. Let's go take a look at some of the things I observed."

At the line, Dave gave both Bill and Norm a pencil and piece of paper and told them to take notes and make some sketches as he pointed out a number of issues. They watched while one of the engine mount assemblers, Robert, tried to attach two nuts to the engine mount studs. It was a poor design, and they had to be put on underneath the engine, which was three feet off the ground. Further complicating things, the engine and its fixture had a variety of protrusions underneath that made it even harder to get at the studs. They watched ten straight units go through. Robert had white tape on three of his knuckles to protect them from getting jammed or scraped. He had to bend over at an awkward angle and twist his back sideways. He reached under with the nut and then worked to get a ratchet around on it. Twice he dropped the nut; four other times, the wrench kept slipping off. It took a lot of time, there was a lot of *variation* in the time it took to complete, and it was an ergonomic nightmare. Norm had already completed a few notes and sketches. They watched Sally for 15 different production cycles. She had to carry a 20-pound pneumatic driver about five feet over to the line, hold it at a three-foot height, and turn in two pairs of bolts at a time in four locations. Sally was a small woman. She adeptly maneuvered the tool, but you could see the strain it put on her. Each time, the work was done within the desired cycle (or takt time) and there were no quality issues, but it was a difficult job. Dave had heard that nobody liked this job. Sally had been doing it for about a month. Bill asked if he could try a few cycles. While it was easier for him, he was still glad he didn't have to do this job every day. He didn't say anything, but he had already come up with an idea that he was eager to try out. He thanked Sally and told her that he felt he could help; she told him that she would appreciate anything he could do.

The three of them moved three stations back upstream and watched as two operators lifted an engine from a line-side storage rack and maneuvered it to the line. Although it was a small engine, it did look heavy, and Dave suggested that Bill and Norm try the next one. They found that the engine was 75 pounds, not a huge burden for the two of them, but it *was* awkward and they wouldn't want to do this all day, every day. Bill wondered why they didn't have a Gorbel crane at this station, as they did at many downstream stations. It seemed like an obvious improvement, and he mentioned it to the group who had crowded around.

Dave took Bill and Norm aside and gave them some guidance. He explained that at his last company, he had some 25 moonshiners in his section alone, and one of the things they spent a lot of time doing was developing devices that allowed operators to easily grab, move, manipulate, and place awkward and heavy components and assemblies onto units on a line, with minimal effort and safety risk. He also said that while cranes were sometimes a desirable solution, one of their goals should be to develop simple, low-cost devices in place of cranes. He went on to explain some of the reasons that cranes weren't desirable on a production line.

NO CRANES

Dave explained that one of his former senseis was so adamant about the need to eliminate cranes and conveyors from Lean lines that Dave himself had become a disciple. He explained that while cranes did have some advantages, there were other ways to accomplish the same goals. He went on to explain that cranes often created conflicts in higher-volume lines operating at faster rates, as they were often shared by operators and one of them was often waiting for the crane. Much time was typically spent moving a crane to a component, hooking it up, moving it to the work-in-process, aligning the part, disconnecting the crane, and moving it back to its home position. Much time was wasted trying to get the part under the crane aligned with the work piece. He explained how much of a safety risk they were and told a few sad stories of horrific injuries when cranes had failed. They also cost a lot of money to install, required a lot of time and money to inspect and maintain, and often had to be replaced not much past depreciation schedules. Finally, once installed, they were not easy to relocate, often becoming "monuments" that people had to work around.

There was one more opportunity on the line that Dave wanted to show the guys before he set them loose to see how they'd do in coming up with improvements to the identified problems. He took them down six stations to observe another challenging process. One of the stations was largely surrounded with a workbench and point-of-use part shelves. One of the parts could not be stored close to the operator due to a lack of space; instead, it was stored behind one of the shelves. The operator had to walk around a four-foot-long shelf, slide in behind it, retrieve a small relay from a box, and walk back around to install it. Dave wondered if there was a way they could adjust the storage racks to eliminate the waste of motion and transportation. Norm took a sheet of paper and drew a few sketches for Bill, who marked up his sketch a bit. Dave let them work through some ideas. He suggested they show what they were thinking to Frank, the operator, and stood back while he observed the three of them interacting animatedly. Once the discussion died down, he stepped in and said,

"Let's go solve ourselves some problems."

Dave stayed out on the line, observing. After just 35 minutes, Norm returned. He was carrying a strange contraption, a four-foot metal rod bent in three places at some weird angles, with a ratchet socket and a small mirror welded on the end. They walked over to Robert and Norm, handed him the device, and asked him to try it. He showed Robert how to load the nut, and then watched Robert stand in a full upright position, maneuver the device under the unit, align the nut with the mirror, and crank the long wrench back and forth. Robert was thrilled, and Dave told Norm that he'd done a great job.

Norm said, "It was pretty simple, really."
"How much did it cost?"
"Nothing. I found all the parts in the back room, just bent them and welded them up."
"Fantastic!"

Dave pulled Norm aside and congratulated him again, but encouraged him to ask Robert if he could improve anything. Robert told him that it was great, but it was a little heavy, and it hit one of the engine mounts when he came around about 45 degrees. Norm asked him to show him again, and got down on his knees and watched the movement.

"No problem. Give me a few minutes."

True to his word, it was not that long until he was back; this time, he'd made the same device out of an aluminum rod he found and had added one more small bend about half-way down. This time Robert used it with no effort or issues. It was a simple fix, not earth shattering, and it probably saved only about 10 seconds per unit. What was important to Robert was that he had perfect posture now when he did the task. He thanked Norm exuberantly. Dave could see that Norm felt great, and he went back to tackle the next issue.

Dave turned and saw Bill rolling out another strange contraption. He'd modified a simple hand truck, adding two more wheels and mounting a pneumatic driver on some sort of slide with a clamp. He rolled it over to Sally, and a few nearby operators took one look at the silly contraption and started laughing. But Sally wasn't laughing. Bill showed her how to roll the truck over to the engine on the line. He adjusted the height of the driver and then clamped it in place. He tried the first one, and with no lifting and no effort, he was able to roll it in and out on all four sets of bolts. He asked Sally to try it. She jumped right in to use it and handled the pneumatic driver with ease. Almost laughing, she said,

"Thank you, thank you, thank you!"
Bill smiled back. "It was really nothing."
"No," Sally replied, "It's ingenious."

The other operators weren't laughing anymore, and gathered around to watch the next cycle. Bill could see that the hose got kinked up as Sally used it. He gathered the hose and wrapped one coil around two metal arms at the top, and it pulled the entire hose up away from the kink points, also eliminating a trip hazard on the floor. Dave congratulated Bill, thinking that even though these were simple solutions, they were "off to the races." Bill and Norm convened at one of the engine assembly stations with a cart containing a number of tools and pieces of sheet metal. In about 30 minutes, they'd crafted and mounted a chute slanted downward so that the relays were conveyed using gravity from behind the point-of-use material storage racks through a shelf on the workbench. It had a small stop at the end to hold a box of relays about two feet from the operator, at about a four-foot height. It was near-perfect ergonomically and virtually eliminated the prior travel and material handling. They had to adjust the chute a few times to ensure that the parts traveled properly. They made sure the edges were bent and any burrs filed down for safety, but in a matter of minutes, they were done. They showed the operator how it worked and asked if they could change anything, but the operator was satisfied with the solution as-is.

In the span of about two and a half hours, Norm and Bill had developed three fairly simple solutions to some challenges in this small section of the line, solutions that made a real difference for struggling operators, and ultimately the business. Their work was the talk of the morning among many operators on the line who had not often seen a lot of regular changes in their areas. There was a buzz growing—a small positive one, but a buzz nonetheless. Bill and Norm's original concerns about interfering with their colleagues and having their motivations challenged were waning. They'd taken Dave's coaching to heart and were listening to the operators, getting their input, and ensuring that the operators were satisfied before handing it off and moving on to a new project. They were starting to like this. They knew they had a creative touch and loved to make things, and seeing their work make such a difference, so quickly, was extremely gratifying … and they were still only in their first day.

Solving the engine lifting problem was quite another thing and was going to take a little more time. But Dave didn't push them to get everything done at light speed. He emphasized not being afraid to take risks and produce multiple iterations of a solution. The keys were to learn along the way and increase their capability to solve more and more complex issues, and to do it right before ultimately handing it off to an operator or team.

Pete wandered onto the floor, curious about what they were up to. Dave led him to Bill and Norm and had them take him through the three devices they'd built that morning. Pete was as impressed by the excitement and positive energy as by the devices themselves. Dave pulled him aside.

"This is just the start. Wait until we really get going on this moonshining stuff."

Dave had some experience with a concept called "7-Ways," another key concept in process improvement and 3P. He didn't know it yet but events were underway and, before long, he too would become a student once again. He had been practicing and teaching what he believed were the real essences of 3P and 7-Ways for quite some time and, in truth, managed to accomplish some great things through others and the coaching he was able to do. He would soon realize he was on the fringe of possibility, as he learned more about the full and pure processes for both, but for now, he knew what he knew and this served him well.

He said to Bill and Norm, "Let's go do a 7-Ways on the lift."

Bill and Norm looked puzzled.

"I'll show you what that is right after lunch. We need another three people to participate. Can you round them up?"
"Sure."

Dave went off to gather a few supplies.

After lunch, Dave was waiting for them with the whiteboard and some flip-charts when five curious and tentative, but interested, employees walked into the conference room. Bill and Norm brought two of their fellow maintenance people with them, along with Gerry Barr, one of the engine assemblers who currently had to do the lift.

Dave started out by saying, "This is really simple. I'm giving each of you a pad of Post-It® notes and a pencil. I want you to think about any way at all to get that engine off the storage cart and onto the frame without any manual exertion or cranes. Be open minded and creative, and don't worry about your drawing skills; just sketch out your ideas in simple forms. When you get a few, go up and post them on this flipchart. Take a look at what others have put up there. It may trigger some ideas. We'll take as long as we need; we want at least seven different ideas up there, but feel free to come up with as many as you can. Any questions?"
One person asked, "Do we limit it only to things we have here currently?"
"Great question, you should have no limits on your thinking. There's no such thing as a bad idea. Sketch whatever comes to your mind that might work. If there are no more questions, let's get started."

After 90 minutes, 14 Post-It® notes had made it onto the flipchart. The team hadn't posted a new one in ten minutes, although most of them still seemed to be thinking.

Dave called a timeout. "Let's see what we have."

They all gathered around the flipchart, and Dave asked each person to describe their idea or ideas to everyone. They all did, to a round of good-hearted chuckles at some of them and some real awe at others.

"We need to narrow down our choices to pick the best. We want something that is safe, simple, fast, reliable, and will do the job and won't cost a lot of money. Why don't each of us put our top three picks in pencil beside each sketch, just put a star beside your picks and we'll see what we end up with."

Each person went up to the flipchart and made their picks, including Dave. When they were done, one of the ideas had six stars, one had five, one had three, and four had one star. They all agreed that Bill and Norm would work on the version that had six stars and the other with five.

About four days later, Bill and Norm came back out to the line. They had tried both designs but preferred one from a simplicity standpoint, and they wheeled out this modified lift jack. It had a simple hydraulic switch to raise and lower a set of parallel arms, and some large casters that rotated very easily. They aligned it to the height of the engine, rolled it in so the arms grabbed underneath two channels in the frame, and then they toggled the up switch a second and rolled the engine away from the cart. They were able to rotate the engine 360 degrees, and roll it four feet over to the frame and slide it over and down. It was a pretty fast process, but getting it aligned on the frame took about 30 seconds, longer than all the other processes, and a small amount of manual effort nudging the engine around on the transfer cart. Dave drew a picture of a cup-in-cone design. There was a natural protrusion of a pin coming off the frame, and Bill and Norm took advantage of this to add a guide cone to the transfer cart and brought it out for the second trial. They also added a number of stop positions on the up-and-down toggle so the operator did not have variations in height each time. This time, the operator who tried it toggled up once and the forks came to the exact height. They toggled once more and it lifted the engine a centimeter. They rolled and turned the engine toward the frame, easily lined the pin on the frame into the cup on the cart and pushed it forward, while it aligned perfectly into place, then toggled down once and it dropped into place.

One of the safety reps came over and expressed some concerns about lift ratings. The team agreed that they would get a reliability engineer who was certified in this area to rate and approve the device, mollifying the safety rep. Clearly, he saw that, compared to the current method, this device eliminated injury risk significantly and, with the rating, would be fully compliant.

Over the next four months, Bill and Norm worked every day, building tools, fixtures, and equipment. They built lifting and conveyance devices, but they also played around with a little more sophisticated equipment. In one instance, they built a roller to unroll paper for masking and reroll it the opposite way on a reel. Sourcing had been able to find a quality lower-cost masking material, but at the moment the only way they could get it was with the paper wound the opposite side up from how they needed it to come off the roll. The savings were significant, and this new roller could change a roll over in four minutes. It was on wheels and was moved to an area right at the paint prep station, allowing one roll to be carried at the line and the next roll to be staged waiting to be changed over. The unit they built was painted and looked professional, as if it had been purchased right off the shelf. In truth, it was homemade with spare parts, built for the cost of labor only, had no bells and whistles, did exactly the job it was intended for—no more, no less—was small and compact, was reliable, and was easy and quick to operate and maintain. Since the start, Pete had been impressed with positive comments from the workforce, whose engagement and buy-in seemed to be increasing as a result. There had been a significant reduction in safety risk and substantial gains in productivity. He'd approved two more full-time moonshiners.

They'd be put to great use shortly and make an even bigger impact on his organization. Month by month, the team continued to surprise everyone with the creativity and simplicity of the solutions they were bringing to bear on the daily struggles the associates on the line—the real value creators—were having.

❅　❅　❅

Dave and Pete sat in the corner bar, celebrating October's off-the-chart financial results. Just that morning Pete had taken a call from Vice Chair Steve Sawyer, who congratulated him on the progress they were making this year. Steve let him know that the CEO was "tickled pink" about the turnaround and was really happy they all had faith and rode it out, but of course was expecting a lot more. Pete had asked him what the CFO, John Cuso, was saying, as John had sent him several nasty notes a few months prior.

"John is John, and doesn't often praise too many people. But he's backed off on pressing for a relocation of your plant and hasn't had anything negative to say in at least eight weeks. For John, that's saying something. Keep at it, Pete. I'm glad we put you in there."

Pete relayed some of the conversation back to Dave.

"So Dave, I really like this moonshining concept. Where do you see us going with it?"
"Well, in the past I've benchmarked other companies that have moved from doing the type of small solutions that we're doing to points where they're designing and building a good portion, if not all, of their equipment themselves. Of course, they didn't get to that point overnight; it took a long time for the team to mature to this level of capability, but it definitely is possible. I'm getting a lot more familiar with our processes and our capital planning, and I think we can make some big changes using the moonshiners and

our manufacturing engineers together toward a different philosophy. First, I looked back over the past three years and saw that we've been spending about $3 million a year on plant and equipment."

Pete interrupted, "Tell me about it. It's getting more and more difficult to get approval for such investments, and frankly I'm not sure that we are getting a good return on them."

Dave continued, "I'm confident we can cut these costs by 40–50% over the next few years, maybe more, so we could save a million to a million and a half in capital expenditures."

"That would be terrific!" Pete said.

"We also have way too much automation and complex machinery. Our Overall Equipment Effectiveness (OEE) ratios are down below 70%. This means that much of our equipment is not available as often as we need to make good parts. We have a lot of monuments, big machines that aren't very flexible or easy to relocate. Our equipment is typically way bigger than it needs to be. A sensei told me a long time ago that a stretch goal is for a machine to be no bigger than twice the size of the part it's working on. Our equipment tends to have a lot of options and features, which we don't use and don't need, and which are difficult and expensive to maintain."

Pete interjected, "I've made suggestions in the past about relocating equipment to improve flow. Then people start telling me what it would cost to move it, and even I have to say 'whoa.'"

Dave went on, "We have no standards for our equipment and many different brands, and it makes for a high learning curve for the operators and makes the servicing more expensive. We have a lot of ergonomic problems out there, and our safety risk is high. Our recordable rate this year is 3.8, not stellar. 'World class' is around 0.7. Also, we have a tough time conveying material and equipment around the plant, and where it *is* easy, we spent way too much money and are very inflexible. I could keep going, but boy, do we have a lot of opportunities."

"Whew!" Pete uttered. "You mean we can do all of that with moonshining?"

"That, and a lot more," Dave responded. Raising his glass, Pete said, "I'll drink to that!"

Chapter 4

Building a Case

"You've got to be kidding!" an exasperated John Cuso, CFO of Enterride, exploded.

The Trail Rider team had made it through the year, with each month in the fourth quarter improving slightly over the one before. They'd returned to profitability in September, and since then, operating margins had been up 35% over each prior year's last quarter. Inventory turns had improved from a dismal 5 to a poor 7.5, a 50% improvement, and were continuing to trend positive. Customers, who for years had either abandoned them or complained about them, were beginning to comment on the improvements in delivery and quality. They'd always preferred the Trail Rider product offerings, but *not* their business performance. As a reflection of this turnaround in customer opinion, orders were up 20% in the fourth quarter. They'd freed up more than 30,000 square feet of production floor space. Perhaps most importantly, they'd gone all year without a lost-time injury, and their rate of recordable safety incidences was down 52%. Pete had conducted their annual employee engagement survey in November, and while the prior year had had the lowest overall satisfaction score ever—a dismal 57%, with many disgruntled employees—this year's score climbed a whopping 14 points to 71%. Obviously there were still many opportunities to improve engagement, but the trends were positive.

Pete, Dave, and their consultant Ken had proven to be a potent combination. They worked well together, their individual strengths complemented each other's, and most importantly, they were all learning. Pete, for one, had never learned so much in his entire career. Despite the performance improvement, they were far from levels of excellence, and they knew it. There were still hundreds of "firefights," or problems, every day. Processes were just not yet that stable and capable of consistently meeting quality requirements. Nobody in the company was naïve enough to think that they were "out of the woods" yet, but it sure felt better than the tough years. The good news was that these incremental improvements, or "kaizens" as the Japanese called it, had them trending upward.

Dave was coaching and influencing the entire division on all the Lean basics that he'd learned long ago. He continued to believe in and spread Toyota's 3P processes. The simulation activity had spread everywhere. As he'd expected, the hourly workforce had quickly seen the advantages of moonshining and began to submit a lot of ideas and requests for help. Most of the moonshining activities continued to focus on small tools, fixtures, equipment enhancement, conveyance devices, and lifting solutions, but the impact was accumulating.

Ken had been coming in to drive kaizen activities every month. He was noticing an increase in involvement from the hourly workforce with each kaizen event. Ken brought a lot of external credibility. Although Dave was recognized as a strong Lean leader, Ken had become his mentor and coach, and attained a high level of respect from much of the team.

❋   ❋   ❋

It was a Friday afternoon at Enterride corporate headquarters, and the CEO, CFO, vice chair, and head of HR were gathered in the executive boardroom, discussing getting a fast start to the new year. They'd just come to a particular item in a project report that Vice Chair Steve Sawyer had submitted. It involved the relocation of an off-road vehicle assembly obtained through a recent acquisition. Steve's plan was to assemble the vehicle in their Trail Rider division in Memphis.

❋   ❋   ❋

John Cuso could barely contain himself. "Apart from the fact they've been one of our lowest-performing plants for years, this is a rare chance to get a fresh start in a low-cost region. Why *wouldn't* we want to take advantage of the chance to get a real edge on our competition for a change?"

Steve responded, "Listen, John, I'm tired of your incessant drive to move everything we have to China and Mexico. If you had your way, we wouldn't have a single job left in the United States or Canada. We've been globalizing for the past decade and we have a lot of great operations in emerging markets, but our strategy has been to localize operations and supply chains around the primary markets they serve. I've supported this, we've made big changes, and it's working for us." He paused for breath. "I'm not negative on moving or growing in emerging markets, just like I'm positive about building and growing operations in the United States or Canada to support market needs here. We've had an agreement that where labor costs as a percent of sales are above 20%, we would work to relocate to low-cost regions. Below 20%, we'd use Lean to drive down costs and improve our competitiveness. We've stayed true to our collective strategy and driven it hard. This is about balance, and I believe we can compete anywhere with Lean, whether it is leaning out an operation in China for China, Turkey for Europe, or Canada for North America. I also believe that from any of these locations, we can export successfully to anywhere else in the world as secondary markets emerge for their products. We're seeing this now, in fact."

"We have a responsibility to our shareholders, and $2 an hour labor allows us to be more competitive than $25 an hour labor. I don't care how you want to spin it," John lashed out at Steve.

"Settle down guys; this is not constructive," said Paula Angle, the Senior VP of Human Resources, "We're on the same team here." CEO Frank Kent looked like he was about to say something, but instead decided to let it play out and listen.

"We've looked at it a hundred ways to Sunday," Steve went on. "I've had a team working with our strategic consultants for a month and a half. We looked at China, and we looked at that new property outside Toluca, Mexico, that we saw a deal on. In the end, it just makes sense to move this down to Memphis. The math is close in all scenarios. If we can get four straight years of 8% productivity improvements, Memphis really jumps to the top of the list."

"Come on, Steve. The Memphis operation has averaged less than 2% productivity improvements over the past five years. Heck, two years ago, they were negative."

"Right. And last year they hit 9% overall improvement and an increase of 20% in labor productivity. One year does not a trend make; I realize that, but sometimes you have to bet on a team."

Steve looked over at his boss and said, "Frank, Paula and I were down there a month ago, and the changes we saw were real. Pete has built a great team. Clearly, things have been on the upswing since he's had some time to get settled in. Their new Lean guy has been tearing up the place, and Pete brought in a Lean consulting company, Kata Movements. Their consultant, Ken, is an ex-Toyota guy. I'm very impressed with him and some of his ideas. He taught me a thing or two, even."

John rolled his eyes.

Steve went on, "There are a lot of attractive reasons to do this apart from the financials. We own the plant, and the original architecture plan allows for a 200,000 square-foot future expansion. We have 14 acres of property there, the plant manager is strong, the labor market there is strong, we have a strong experienced workforce, we're close to our major customer base, and 60% of our supply base is within a two-day drive of the plant."

John appealed to Frank. "I still think that there's a lot of brawn going on down there, and it's going to be tough for them to continue and sustain their pace of change. This opportunity is just too great to reposition our cost structures in this company. Our shareholders are looking for stronger growth in operating income. We have to step up and deliver."

Paula Angle spoke up again, "Frank, like Steve said, we took a trip down there last month. In my estimation, what they have going on there is real. I did a number of focus groups with salaried and hourly employees, and heard some common stories. Pete is really changing the culture in the organization. They had a big improvement in their engagement score last year, by far the best in the company. You can't fake that. The employees are speaking. There's something positive going on down there, and I think we should reward them. Let's give them some more business. This team will run with it and do us proud."

Frank responded, "Thanks, Paula. John, you have some valid points, and we shouldn't just ignore them. We do have to grow operating income substantially the next three years to justify the multiples we're getting on stock price and push them higher. But I've looked at the financials of the plan to move to Memphis, and they're strong. They meet our target thresholds, and I agree with Steve's other list of intangibles—in particular, his point that we chose a strategy, it's working for us, the market is rewarding us, customer opinion scores have climbed the past three years, and this is not the time or the opportunity to go off the ranch. So you both really liked what you saw down there?"

Steve jumped in, "Frank, they were even doing this 3P stuff down there!"

"3 what?" mocked John.

"3P, Production Preparation Process. It's a Toyota tool. Their Lean guy, Dave, walked me through it on the floor. Many of the changes they implemented made use of it. Pretty cool stuff."

Paula added, "People mentioned it in several of our focus groups, so I asked some questions. The process really engages the associates to identify relatively quick wins in terms of quality, productivity, and safety improvements. Most importantly, the associates genuinely feel that the changes made are to help them rather than something being done *to* them. I know that I want to learn more about it for possible application to our other businesses."

Frank nodded his head, and sought to bring this item to a close and move on to more pressing items on the agenda.

"Last year things weren't looking very good down in Memphis. What got us to the dance was betting on and supporting our people and implementing sound business strategies. I'm not about to change that. Let's make the move, but Steve, we need to nail this one; no misses."

"Okay, Frank. Thanks. We should also think about expanding 3P throughout the company as Paula suggests. It can really be a game changer."

Frank surveyed him across the table.

"That's a pretty big testimonial. Why don't you find some time on my calendar in the next month and walk me through it. We need a few more game changers around here if we want to repeat another great year. 3P. Huh."

❋ ❋ ❋

It was Saturday afternoon, and Pete, Dave, and Ken were with their wives and kids at Pete's house. It was an unseasonably warm winter day in Memphis, and the wives were gathered around the table on the patio listening to Debbie, who was describing her recent business trip. Ken looked over at his wife, reflecting on the events of the past few months. Julie looked back and smiled. Just last night, she had commented on how much happier he'd seemed recently.

Their kids were running around in the backyard, getting along as though they'd known each other their whole lives. Pete reached into the cooler and pulled out three more ice-cold beers, twisting off the tops and handing them over to his colleagues. He raised a beer toward them and told them how much he appreciated all they'd done for the plant and for him.

"Just last March, I was worried we wouldn't be able to turn things around. Corporate was breathing down our necks, especially John Cuso. And now here we are. With your help, we closed a great year. I'm looking forward to the new year."

Pete asked everyone to gather around. He had an announcement to make, and he topped off their drinks before proposing a toast.

"To great friends and a great year, personally and professionally."

"Hear, hear," echoed around the patio.

"I have a big announcement," Pete continued. "I got a call from Steve Sawyer yesterday afternoon. They're bringing the Trail Gripper product line over to us as they spin off parts of that new acquisition. It'll complement our Trail Rider and Traction lines, and it means that we'll be expanding the building, hiring new people, and driving Lean into everything we are doing. Ken, it looks like we'll need more of your time."

Debbie knew how much this meant to her husband. She'd watched him struggle through the first half of this year, and he deserved this.

Dave said, "Jeez, Pete. You never told me I'd have to work harder than ever when I took this job."

Pete responded, "You better believe it. Now we really get to show our stuff. I can't wait to tell the whole team. They all deserve it."

The friends, their wives, and their kids enjoyed the rest of the lazy afternoon. Things were looking up.

❋ ❋ ❋

It was Monday afternoon, near the end of a tiring but inspiring day. Pete had assembled a team of his best people to begin planning the integration of the Trail Gripper vehicle assembly process into the Memphis facility. He'd just ended a very successful all-employee meeting before lunch and had given what might have been the best motivational speech of his career. There were no signs of negativity. The employees, already on a high coming off a good year after a long

Table 4.1 Project Trail Gripper Objectives

Objective	Target
Project Completion Date	27-Feb
Project Capital Expenditure	$42,000,000
Project Expense	$2,500,000
Product Cost	–3%
Initial Quality (ppm)	1000
Initial Delivery (lead time weeks)	4
Initial Delivery (% on-time)	96%
Inventory Turns per Year	12
Target Volume per Year	40,000

period of being the pariahs of the business, were absolutely pumped. Choruses and shouts of "Yeah!" and "Let's do it!" had echoed through the cafeteria, where the meeting had been held. Lunchtime had not stunted the enthusiasm. The atmosphere was electric.

Pete pulled together Dave and Ken, Mike Young (the shop manager), Bill Stark (the material manager), Lou Marks (the quality manager), George Hall (the AME leader), Lean leaders John Lee and Gina Nelson, and Mary Long (the inside sales manager). This was a good chunk of his team, which he felt would really help to put a plan together to execute this project and meet the objectives that Steve Sawyer had sent (see Table 4.1).

"I got a call on Friday afternoon from corporate—Steve Sawyer and a few of his staff members in particular. He thanked us for a great turnaround this past year, and made it clear that we'd earned the confidence of the CEO. I can tell you, he was speaking to all of you as much as to me. We wouldn't be here without the folks in this room."

Pete continued. "As some of you know, Enterride acquired a new business last June and has been working to integrate parts of it, restructure parts, and spin off the rest. One of their business units made an off-highway vehicle they called the Trail Gripper (see Exhibit 4.1). We never really competed much with it, nor did they have a huge market share. This was never a strategic line for them; they themselves picked it up in an acquisition. It was a model geared around a niche market, aimed at the lower end of the market and targeted to the 15–25 year-old segment. Steve said that the commercial team likes the line and feels it can be a great complement to our base of products here, filling a gap in our existing product offerings. They want to invest heavily in growing market share and are going to put a ton of resources and money into this. They want us to build it here.

Exhibit 4.1 Trail Gripper.

"Steve said they had a preliminary team put together a high-level set of targets to build a case for investment in the new line. We don't have a lot of detail on how we'd manufacture them. That's up to us. They based the targets on growing market share through improved delivery times and slightly lower pricing. Of course, this means we have to lower product cost in order to maintain an acceptable profit margin. As you can see from the chart on the screen, we have a little over a year to expand the facility, develop a process, and start production. We have a pretty good investment target. I was surprised at how high it is, but Steve said they toured the existing plant and there was some very expensive automated equipment. They've budgeted $42 million in capital and $2.5 million in expense for the project. They've given us targets to reduce the existing Cost of Goods Sold (COGS) level by 3%, and to get external quality to 1,000 parts per million (ppm). I think they're currently at about 3,500 ppm, so we have some work to do. We need some pretty strong service performance out of the gate, and they have challenged us to start with 12 inventory turns per year. I'm a little worried about that goal. While we are making progress in our current product lines with the speed in which we turn inventory into sold finished product, we're still under ten turns per year. The commercial team thinks they'll sell 40,000 a year within the first two years of start-up, and have future growth projections that they didn't share with me yet. They also sent along a preliminary set of financial targets, which I've put up on the screen (see Table 4.2). You can see they want us to target a 15% operating income objective. This is a big push from where we are today. I don't have a lot more information than that yet, although I have some pictures and some engineering specifications I'll hand out. We'll learn a lot more over the next two weeks. There's

Table 4.2   Project Trail Gripper Financial Estimates (Year 1)

Measurement	Individual	Year 1 Targets
Product Price	$6,500	$6,500
Volume		40,000
Revenue		$260,000,000
Product Cost	$4,550	$182,000,000
Direct Material Cost	$3,770	$150,800,000
Labor Cost	$390	$15,600,000
Variable Overhead Cost	$390	$15,600,000
Gross Margin	$1,950	$78,000,000
SG & A Cost		$39,000,000
Operating Income		$39,000,000
Gross Margin	30%	
Material Margin	58%	
Labor Margin	6%	
Variable Overhead Margin	6%	
SG & A %	15%	
Operating Income Margin	15%	

nothing I've seen that should be a major problem for us. This is a huge opportunity."

"It sure is," said Mike Young.

"Does anyone have any initial thoughts?"

"If I could," said Ken, "I'd like to throw out a suggestion."

"Absolutely," Pete said.

"In my six months working with you, I've been impressed with how much you're learning, and how willing you are to stretch and try things. This is a very unique opportunity to do a 'greenfield' startup, to design the building expansion and the process any way we want, as long as we meet these parameters. In Toyota they have a process called 3P, the Production Preparation Process, and it's a perfect fit for this kind of project."

Dave chimed in, "I agree. This is exactly what we've been doing this past year."

"No disrespect intended, Dave, but what you guys have been doing are *elements* of 3P. They're good elements but I'd like to propose that we complete this whole process from soup to nuts."

Dave was a little surprised. He'd thought he was an expert in 3P.

"Are you saying that what we've been doing isn't 3P?" Dave asked.

"What I'm saying is that there's a very prescriptive process for 3P that is intended to lead to what Toyota calls a 'Vertical Start-up.' If followed properly, the 3P process will ensure that we meet the schedule, cost, and quality targets virtually right out of the gate."

Pete leaned forward a little more and listened intensely. They all did.

Ken continued, "It would require a little more up-front investment in money and resources than is typical, but that's precisely the principle. We want to be much more proactive and get everything designed, built, tested, and proven out before we ever start. We'll develop a process built with total Lean thinking right from the beginning, in all regards. We'll have full standard work at launch. The extra early costs will more than pay back and quickly. I took the spreadsheet you sent me, Pete, and entered some quick thoughts on the differences we could expect. If I send them to you, can you put them up on the screen?" (See Tables 4.3 and 4.4.)

"You can see that I made some quick assumptions. We can easily save 20%, or almost $8.5 million, on the capital expenditure targets. That alone will justify the extra costs of around $2.6 million to drive 3P the right way. Keep in mind that these are quick estimates, but I feel directionally good about all of them. We should be able to reduce target product cost another two points, and get them down 5%, for an additional $3.64 million in annual savings. There's no question we can do better on initial quality and take inventory turns up to 20."

Dave whistled. Pete looked skeptical. He wasn't the only one.

Table 4.3 Trail Gripper Objectives

Objective	Target	Improvement
Project Completion Date	27-Feb	
Project CapEx	$42,000,000	−20%
Project Expense	$2,500,000	
Product Cost	−3%	−2%
Initial Quality (ppm)	1000	−500
Initial Delivery (lead time)	4	
Initial Delivery (on-time)	96	
Inventory Turns	12	+8
Target Volume per Year	40,000	

Table 4.4 Project Cost/Savings Estimate

Objective	Estimate
CapEx Savings	$8,400,000
Product Cost (Unit)	$91
Product Cost (Annual)	$3,640,000
Year 1 Savings	$12,040,000
Kata Movement Cost	$900,000
Salaries	$1,440,000
Other Costs	$280,000
Total 3P Costs	$2,620,000
Payback Years	0.2

Pete finally said, "I don't really know anything about the full 3P process you're suggesting, but I've trusted your coaching every step of the way. Do you really think we can hit these targets, Ken?"

"Yes, I do, Pete."

Pete thought about it and said, "Well…it's a two-month payback, if I can get corporate to agree to spend more up front. If we get a quarter of what you think, it would be worth it. Listen, team, we're over our time here. I just wanted to give you some background on the project. I'm going to call a series of meetings with everyone over the next two weeks, and we'll put together the plan. I'm not sure what you all think, but I feel we should also throw Ken's suggestion into the mix and consider everything. I'd like all of you to think about how we should approach this and bring in some thoughts tomorrow at the same time, okay?"

Everyone nodded and agreed. They were excited about the full 3P process and wondered how it'd differ from what Dave had been teaching them.

As they were walking out, Pete pulled Dave and Ken aside and asked if they could meet after work for an hour over a few drinks. He wanted to kick around the 3P proposal.

About four hours later, the three of them were sitting in the corner of Kerrigan's Pub, each of them nursing a tall glass of Guinness. The conversation went on for over two hours, and Ken described as best he could the essence of what the process would be. The other two asked a lot of questions. They began to realize that what they'd been doing to date were really simple, incremental improvements to an existing production line. The full 3P process involved much more, and involved radical or breakthrough improvements. They would have the opportunity to design the new line and all its elements from scratch. Ken explained that there would be opportunities to redesign the existing product, which would result in significant cost savings, both in materials and labor. Their excitement, confidence, and willingness grew by the minute. By the end of the conversation, Pete was sold.

"Listen, this is a huge risk, but I want us to knock this out of the park, and I like what I hear. We haven't had a great track record at launching major projects. Unless we change how we do it, we likely can't expect much different, and I don't want to let Steve down. Dave, what do you think?"

"Everything we talked about makes sense. I like it, and I want to learn more about this. Let's do it."

"I agree," said Pete, and with a final clink of their glasses and a downing of the last few fingers of beer, they wrapped it up, heading home to their families.

Although none of them could know how it would turn out, they'd made the case. The journey to breakthrough began!

Chapter 5

Preparing for the Preparing

"Welcome, everyone," Ken Saguchi boomed from the front of the state-of-the-art training room that Pete had constructed last year in his commitment to quadruple classroom training for all employees in his plant. The room had whiteboards across three walls, two breakout rooms in the back, a 52-inch high-definition television with DVD player, an overhead projector, flexible adaptable tables on wheels, and comfortable chairs.

"You're about to embark on a new journey that will transform your company forever."

✳   ✳   ✳

Two weeks earlier, Ken had met with Pete and Dave to start preparations for the 3P work they had all committed to. They were sitting in a spare conference room at the side of the plant that would serve as mission control for their project. The room was dirty and still buried under a lot of clutter, but an almost 40,000-square-foot section of plant floor space had been emptied of inventory in the past six months, a by-product of continuous kaizen efforts. Ken had insisted on meeting here rather than in the comfort of Pete's office.

Ken started things off.

"We need to lead by example and clean this room up if we're going to use this as mission control for the 3P project. This has to exemplify the principles of 5S that we need to build into the entire new process design."

Dave started laughing.

"What's so funny?" Pete asked. He knew that laugh and suspected it was at his expense.

"We need to start with Pete's office first. Then we can tackle this room."

Ken was laughing now too.

"Very funny. My office isn't that bad." Ken and Dave both raised their eyebrows. "All right. I'll show you guys. I'll take on organizing my office *and* lead the 5S effort in this room."

"Ken, what's your schedule like? We may need you to coach Pete full time."

When the laughter died down, Ken said, "So listen, we need to think through the team."

"Let's not short-change anything, Pete interjected. "Steve called me the other night. He wants this done right, even if we have to stretch the budget a bit."

Ken replied, "I'll ensure that we have the right number of people. But let's keep in mind that this isn't about spending money. It's about saving money. We'll put in an initial up-front investment, but we'll beat the targets from corporate no matter what resources we put into this."

Ken asked Dave for a few sheets of butcher paper and taped them together against a whiteboard.

"Why not just type it up on my laptop?" Dave asked. "We can save it and access it later."

"We could just write on the whiteboards," Pete suggested.

Ken took a second and then imparted another important lesson to his favored colleagues:

MANUAL, VISUAL, AND FLEXIBLE IS GOOD

I've found that the key principles of Lean should and do apply to all aspects of work. We want to make sure as we drive our 3P project that all key information is always highly visible to all members of the team. Entering some of our information into a computer is not always a bad thing, in some cases it will make sense, but where we can, sticking to manual recording of information creates a more intimate ownership for the team, and a higher level of commitment to and knowledge of why it's there and what it means. We do want to ensure that the information exhibits remain flexible. During the project we'll continuously be moving around, adding to the information, and we may want to bring information to other areas where project work is being done. We should try and put most of our information on portable media. If it's in a computer, we should always print out in large-size format and publicly display it; if manual, we should write it large on separate paper.

Ken started to draw a grid pattern on the butcher paper. Pete and Dave watched him closely, not sure what he was doing at first. As he added labels, they could see he was developing a project staffing chart. In the left column, he was adding what looked like the different phases of the project, and across the top he was adding what appeared to be roles. After putting down his pen, Ken explained the importance of continuity, and that he would need many staff members on the project fulltime. There were multiple advantages to keeping individuals in full-time roles for the duration of the project, Ken explained, particularly because they would gain a deep understanding of all aspects of their roles and the entire process.

Ken continued, "In the past, organizations sometimes had a hard time keeping a role staffed with the same person throughout the life of the project, and this would be manageable as long as it was only in a few cases. A lot of the work would be simulation, observation, and kaizen. If a few of the roles had different folks rotating in and out, as long as the role was always staffed, it would be fine. Several roles would be part time. There are certain phases of the project in which they would need the additional help, and some phases they would not."

Pete asked Ken to fill in the chart as he saw fit (see Table 5.1). When Pete looked at it, he whistled and commented,

"Wow, that many?" And he went on, "I know I was clear that we wanted to do it right, but I didn't think it would require that many people."

Ken said, "I've walked your current processes for quite a while now, and while this is a new product line, many of the processes will be similar. This is a pretty large scope; we have a lot of processes to design, a lot of equipment, fixtures, tools, and conveyance devices to design, and a pretty short time window. What I've put up here is pretty typical to do this right and ensure a successful start-up. I can assure you again that this is just an up-front investment, and the payback will be significant."

Dave said, "You've listed a product design engineer. What will his or her role in the design of the production process be?"

Ken responded, "I understand your question. Most of the changes you and your team have made so far on the existing line have involved methods improvements, better organizing the workstations, and the like. True 3P involves the product design as well as the process design. For example, we'll want to address manufacturability issues up-front, not later. This means that elements of the product design may need to be changed. There may be other opportunities to hit our target cost through product design changes as well."

"This really is different from what we've been doing," Dave said. "I just don't know if Engineering is willing to participate as a full-time team member."

Pete came back, "The company is placing a big bet on us and this product transfer. We have unintentionally botched most major programs we've done here the past five years or more, and I'm committed to doing this right. Let's not short-change ourselves. I'll fight for what we need; and if we get scaled back, we can cross that bridge when we come to it."

"I agree," said Dave.

Ken watched the exchange with satisfaction. The best 3P processes included a lot of elements, but faith and commitment from the senior leadership team was one of the most important ones.

Ken said, "Let's put the product design engineer on the list as a part-time team member for now. Pete, this may help you in your discussion with the manager of Design Engineering. I think in time they'll see the valuable contribution that they can make through their participation on the 3P team."

Table 5.1 Project Trail Gripper 3P Staffing Plan

	Production Flow Team							Material Flow Team				
Process Plan	Adv Mfg Eng	Lean Leader	Line Supervisor	Quality	Product Design Eng	Moonshiner	Management	Process Plan	Sourcing	Material Planning	Order Entry/Master Scheduler	Material Control
#	2	2	2	1	1	3	1		1	1	1	2
Time	Full	Full	Part	Full	Part	Full	Part		Part	Full	Full	Full
2D Process Design	O	O	O	O	O	–	O		–	–	–	–
Prepare Process at a Glance (7 ways)	O	O	O	O	O	O	O		–	–	–	–
Yamazumi Chart/Std Work	O	O	O	–	–	O	O	Prepare Material and Information Flow Chart	O	O	O	O
Small Mockup (Layout)	O	O	O	O	O	O	O	Small Mockup (Layout)	O	O	O	O
Process Evaluations	O	O	O	O	–	O	O	Process Evaluations	O	O	O	O
Full size Mockup (Layout)	O	O	–	O	O	O	O	Prepare Receiving and Shipping Workflow (Standard Work)	O	O	O	O
Order Equipment	O	O	–	–	–	O	O		–	–	–	–
Production Trials	O	O	O	O	–	O	O	Production Trials	O	O	O	O

12	Total Full Time
6	Total Part Time

Legend:		
	O	Require
	–	Not Require
	O	Maybe require

They could see that they needed 12 full-time team members and 6 part-time contributors. They started to throw together a team list, adding, subtracting, and revising over and over as they brainstormed the best team with the best chemistry they could find. Pete knew he would have to backfill many of the roles, and he also knew that he needed to take some of his best people and assign them to this big critical initiative. He also knew that he needed to be thoughtful and ensure there was a good plan to not significantly weaken the current operation or its continuous improvement. Not all leaders were comfortable with these kinds of decisions; they often were wedded to current good leaders in their current roles and were just too nervous about moving people around. Pete was a good leader; he had never done 3P but he had taken on many large initiatives in his career right in the middle of ongoing operations, and he had always had the courage to make the right decisions regarding people. He was determined once again to do the right thing for both this project and the existing operation, and he knew they would find a way. After a couple hours of thinking this through, and without yet having a full backfill plan, the three had pulled together the first draft of what they felt would be an exciting team (shown in Table 5.2).

Many team members had been involved heavily in the Lean journey over the past year, with a few others who had not been involved that much to date. They decided they wanted to have an hourly operator on the team and selected Johnny Cox to represent the assembly associates' needs and thinking in the process design. Ken had mentioned that while this was a comprehensive list, there would be many opportunities to pull in other hourly operators to help with development. He did tell them that many operators would be fully trained on the process before they even launched. In a stroke of luck, they discovered that two of Pete's best employees—Mary Long, the current inside sales manager, had a prior background as a master scheduler in another Enterride factory; and Sylvia Bennett, who had been a material planner prior to her order entry role.

Pete requested Ken that, for now, they not post the team list until they'd had a chance to shore up the whole staffing plan. He didn't want the rumor mill to blow up and cause anxiety and hurt feelings and much uncertainty until they had it finalized and had in place an effective communication plan to all affected employees. Ken totally agreed. He understood that respect for people was an overarching principle. They would post the team list front and center when ready, but for now, until finalized and communicated well, this was viewed as sensitive information.

Pete asked what was next. Ken told him they had a lot of planning work to do as well as getting on rapidly with the execution, but he preferred to do most of it right from the start with the total team, to get all of their views, ensure a stronger plan based on many diverse and committed inputs, and a plan that the entire team felt a great deal of ownership for. Pete went to work with his staff and Steve Sawyer, and together they were able to develop a plan to support the staffing model they'd developed. They had to move several people around to backfill with other expertise within the corporation and hire a few new people. This is how the team found themselves sitting together on the first day of the

Table 5.2   Project Trail Gripper Team List

3P Role	Expertise	Team Member	Current Role
Project Manager	Leadership	Dave Martin	Site Lean Leader
Production Flow Design Member	Lean	John Lee	Lean Leader
Production Flow Design Member	Lean	Gina Nelson	Lean Leader
Production Flow Design Member	Direct Supervision	Mike Young	Shop Manager
Production Flow Design Member	Assembly	Johnny Cox	Assembler
Production Flow Design Member	Manufacturing Engineering	George Hall	Manufacturing Engineer
Production Flow Design Member	Manufacturing Engineering	Harry Givens	Maintenance Manager
Production Flow Design Member	Quality	Lou Marks	Quality Manager
Production Flow Design Member	Product Design Details	Brenda Lewis	Product Design Engineer
Equipment Design Build Member	Mechanical Building	Bill Cook	Moonshiner
Equipment Design Build Member	Mechanical Building	Norm Wilson	Moonshiner
Equipment Design Build Member	Mechanical Building	Byron Hill	Moonshiner
Material/Info Flow Design Member	Supplier Selection/ Negotiation	Betty King	Commodity Leader
Material/Info Flow Design Member	Material Planning	Sylvia Bennett	Order Entry
Material/Info Flow Design Member	Order Entry & Master Scheduling	Mary Long	Inside Sales Manager
Material/Info Flow Design Member	Material Control	Bill Stark	Material Manager
Material/Info How Design Member	Material Control	Scott Green	Material Controller
Material/Info Flow Design Member	Supplier Processes and Quality	Linda Campbell	Supplier Quality Engineer

project, in Pete's brand-new, high-tech training room, about to embark on the journey of their lives.

❊　❊　❊

"Let's get started." Ken Saguchi, the 3P sensei, stood at the front of the room in front of a flipchart, ready to walk the team through a four-hour training session using a PowerPoint® training module and handwritten flipchart notes.

"Who knows what 3P is?"
George Hall shouted, "It's about having a vertical start-up," obviously having listened to some of Ken's coaching early on.
"That's right, George. It's about a vertical start-up, where we develop a process that will launch into full production, exactly on the target launch date, at target production rates, with target quality, and meeting target costs. In other words, we turn the key and go. How often have you hit these targets at launch date?"

A chorus of laughter rippled around the room.

"It's also about focusing on the hardware and software elements of the process, from hardware like equipment, fixtures, tools, and materials, to software, like information flow, organization, standard work, and inspection systems."
"Why would we do 3P?"

Several hands shot up.

Ken pointed at Mary Long. "We do it to develop world class processes up-front, before we even launch the new process."
"That's one reason. Mike?"
"It's a competitive weapon that can give us a real advantage in our industry.

From the back of the room, Sylvia Bennett slowly raised her hand and quietly added that it helped to promote rapid continuous improvement.

"You're all right on. Great answers.
"There are six major reasons that we would want to do 3P up-front: a new plant start-up, a new product development project, a product transfer from one plant to another, and in support of a major design change to an existing product. Can anyone guess what the two other reasons might be?"
Gina Nelson was quick on the draw. "For major plant rearrangements and major changes in demand."

"Very good, Gina. You nailed it. So, there you have it folks; these are the six conditions that should make us consider doing 3P up-front. By the way, through this we want to ensure that we build appropriate quality into the product and the production process; that we create highly flexible and adaptable equipment and process to enable simple, fast, easy, and low-cost continuous improvement in the future; and that we deliver exactly on schedule at the target cost points.

"The old way of developing operational processes is to first develop a plan for equipment and machines, then to develop the tools, and finally to develop the layout and flow. It's how most projects are sequenced and prioritized. In our 3P work, we're actually going to flip the order of the activities. First, we'll focus on developing the process and flow, then we'll develop the tools, and only then will we focus on developing the equipment and machines. These will be designed to fit a Lean process design using Lean tools. Any questions?"

Nobody raised their hand. Ken surveyed the crowd, trying to determine whether people really understood or were struggling and afraid to comment. It seemed from their body language that the team members were getting it, were highly interested in it, and were even getting excited by the training so far. He was impressed that Pete was in the room and had committed to taking the training with everyone else. It sent a strong message to everyone.

Ken then spent the next two hours or so on two major topics before the team would get heavily involved in 3P design. These were the principles that they would base all their design activities on, and he needed the team to really understand them. The first was a series of slides showing 17 of the "Do's and Don'ts of Manufacturing," as shown in Exhibit 5.1. The other slide showed 20 of the principles of world-class manufacturing. It was Ken's intention to coach the team intensely to avoid the "Don'ts of Manufacturing" and embrace the "20 Principles of Manufacturing." When they built their Mission Control room, Ken would have both of these exhibits posted on large posters to make sure the team continuously referenced them throughout the project.

"Let's look at the 'Don'ts' and alternatives for each. We need to avoid the 'Don'ts' and practice the 'Do's,' as one might call them, whenever possible."

He warned the team that some of these would be tough because they were probably the exact opposite of how we had tended to design our processes for the past 20 years or more. He took his time explaining each one, wanting to make sure that the team fully understood each. Ken gave the team time and encouragement to ask for more clarification, even to challenge. There were some lively debates on some of them, although mostly it was a positive discussion. The team wasn't shy. They asked a lot of questions and pushed back a lot, but in the end, everyone in the room understood and was willing to embrace the Do's and Don'ts, supported by Ken's coaching.

Exhibit 5.1   Dos and don'ts of manufacturing.

"We need to be clear: the goal is to start with raw material and focus on all the process steps in between to fulfill customer needs with the right value stream proposition. One of our goals is to ensure that the information, the material, and the work in process flows throughout, and that each process step adds value, with minimal non-value-added activities in between the steps."

Everyone understood. This is what Dave had coached them on for much of the prior year. Ken pointed out this fact to everyone and made sure they understood that they wanted to take the creation of value and the elimination of non-value-added waste to a whole new level.

Reviewing the "Don'ts," he explained that they needed to get to one-piece flow through the whole process and work hard to ensure that work didn't move in batches. He emphasized the impact on material and work-in-process of "pulling" it through the process versus pushing it. He explained to the team why this was important, reinforcing Dave's prior coaching. In fact, throughout this training, the team constantly referred to one thing or another that Dave had coached them on.

Ken launched the rallying cry for the next year. The team would get sick of it at times, but this one was non-negotiable, Ken proclaimed:

WE HAVE NO MONEY

Folks, we need to approach all our development work from the standpoint that we have no money to do any of it. We need to get creative and reach for "ideas, not money." We always need to look to the simple, the small, the manual. Keep it simple. Of course, this literally doesn't mean that we'll spend no money at all on capital, but we'll always try to find ways to spend as minimally as possible. There are many examples of companies where creative teams engaged in 3P developed processes at a fraction of the cost they typically would have. Remember all the time, as you do your design work, our collective rallying cry: "We have no money."

Ken continued, "We'll develop right-sized, simple, flexible, and safe-to-operate machines. We don't need operators 'machine watching' or standing by as a machine is running. This is a horrible waste of people's time and abilities. We need an environment that ensures that everyone is fully able to participate in the process and that the work will flow. As we said earlier, our goal is to create processes that enable one-piece flow. Unfortunately, a lot of times the equipment we use between work steps is capable of handling many pieces of material. This encourages operators to pile more than desired between steps. Examples are long conveyors, shelves in between stations, big carts, and stacks of pallets, all of which can hold many parts. If we develop conveyance methods that limit material, operators no longer have the option to add more quantity than standard and desired.

"Visual management is very important in a workplace. We want people to be better able to see the flow, or the absence of it, and to easily spot abnormalities so that they can deal with them. If we restrict the height of things like shelves, benches, and equipment, then we create an environment that makes it easier for people to see, and one that is safer as well. It is very common for manufacturing engineers to specify and develop very large equipment that is much bigger than it needs to be for the tasks it does. This equipment is often expensive and takes up space. It usually represents a 'monument' that is difficult to move, is inflexible, and is difficult to maintain and update. Sometimes it's difficult to believe that a task can be done with a device that is significantly smaller, but with 3P thinking, this is made possible. Some even say that ideally a piece of equipment should be no larger than twice the size of the part it's working. You may scoff, but this is as good a stretch goal as any other. We also don't need machines with many options or 'bells and whistles' that few will use. Think about the products you buy for your home. Sometimes the salesperson sells you on the one with a lot of options and features, yet most people only use the basic features. It's no different in the work world.

"During the 3P process, we encourage you not to agonize over choices and barriers. Your mindset should be 'just do it' or 'just try it.' If we are successful with this, we'll also set the tone that the same thinking should translate into the actual work environment when we complete the project. Often during a change effort we find people who are close-minded and inflexible. You know what I'm talking about. Change can be scary, and for most people working outside their comfort zone is unsettling. Yet, in 3P, we absolutely need open-mindedness and fast-paced decision making. Adulthood has a way of squelching people's creativity and can limit their potential that can be unlocked with 3P. One of the most important lessons that will be stressed over and over again throughout the 3P exercise, and which we need all of you to fully embrace, is the following."

THINK LIKE A 12-YEAR-OLD

I want you to think back to what life was like when you were 12 years old. Remember? Remember how much was possible? It seemed you could do anything. This is perhaps one of the most important lessons to learn and to stick to when doing 3P. We often become jaded as adults; barriers and obstacles seem bigger and bigger, and more and more things feel difficult to do, if not impossible. When we were 12 years old, we could do anything we wanted and, perhaps more importantly, we were often at our most creative at that age. We want you to bring out your full creative self during this work. We want you to be fully open to possibility. We want you to just believe anything is possible. We want you to think like a 12-year-old.

"Too often, when we look for a solution, we look to a catalog. We go online and search for that solution. Instead, I want you to look within yourself for the simple answers and to develop your own solutions. With 3P, we'll push people to get their hands dirty and to do '7-Ways.' Don't just think of one way to do something. Come up with seven. Really push the creative envelope. We'll also encourage the concept of 'trystorming' versus brainstorming. While the latter can be a good process to get folks open to more and more ideas and not be close-minded, it doesn't go far enough. Simply talking about things doesn't create solutions or expose whether or not an idea is a good one. We'll encourage everyone in 3P to try out his or her ideas, to simulate them, to make them more real. It's in the trying that we will make real progress and find out what works and what doesn't work. Most importantly, you'll discover new ways to enhance the ideas and make them better.

"If we look at our typical factories, they look much like suburban neighborhoods with separate family dwellings. Machines are spread out and separated from each other with lots of space in between. This means that we have to move material a lot farther between them. We risk injury, and we encourage the buildup of inventory when this happens. We should think about designing factories like townhouses. Keep the machines tightly together. Don't allow any space between them, and pass material from machine to machine. We also want to build flexible factories. We'll encourage—in fact, we'll *demand*—continuous improvement. The easier, faster, and cheaper we make it to try a change, any change, the more likely people are to do a lot of them over time. We limit continuous improvement when we create inflexible, difficult-to-change 'monuments.' Let's put our equipment on wheels when we can, make it small and light, and use quick disconnect methods rather than 'hard' utility connections. With 3P, we'll also encourage simple, creative mechanical thinking. Handling parts in and out of equipment can be a repetitive, time-consuming task that adds no value and increases safety risk. Creative teams have learned to develop simple, low-cost mechanical methods to move parts in and out of and between machines. This is done in the spirit of reducing waste and enabling one-piece flow. We'll do the same."

During the six months he'd spent at the plant, Ken had noticed that no matter how they approached the 3P work they would do, that cost would keep coming up front and center in their minds.

Ken commented on this. "I know that it will be difficult, but I want you to not worry so much about cost. You should only think about maximizing quality performance, meeting the takt time objective from the start of production, and nailing the launch date. I can assure you that cost will take care of itself."

Ken paused, took a deep breath, and looked out over the team in the classroom one by one.

"Much of the work that you do in 3P will be to minimize the degree of the 'Don'ts' and to maximize the degree of the 'Do's.' For whatever reason, most of our factories resemble the 'Don'ts' side of the lessons; just look around when you go back out into the factory. If we don't intentionally keep all these lessons in mind all the time as we develop our new process, we'll tend to lean toward the 'Don't' side, and we can't go down that road. This is critical to developing the best process we can. I'll coach you, but all of us need to really work hard on this throughout the program.

"So, now that we've spent some time reviewing the 'Do's and Don'ts of Manufacturing,' let's look at a separate list that I got from one of my own senseis years ago. He called it the '20 Principles of World-Class Manufacturing' (shown in Table 5.3). Several of these are duplicated in the 'Do's and Don'ts of Manufacturing,' but I want to walk you through some of the items on the list. With 3P, we'll work very hard to adhere to all these principles. Like the 'Don'ts,' many of these are counterintuitive to what we would find in traditional manufacturing plants. They're not always easy to

Table 5.3 Twenty Principles of World-Class Manufacturing

It is critical to our success in achieving our goals that we not compromise these principles or ask others to.	
1.	One-piece flow
2.	No forklifts, no cranes, no hoists
3.	Mistake-proofing, go/no-go gages in use
4.	Product and operators do leave the cell or line
5.	No process reversals
6.	Standard Work Combination sheets in place
7.	Systems and standard work in place to prevent parts shortages
8.	Cross-training is actively done; everyone can do everything
9.	No trash containers
10.	Moving line, used as a pacemaker for the process
11.	Daily checks and TPM are carried out routinely, no machine downtime, time deducted from time available
12.	Quality Control Process Charts (QCPCs) will be routinely used to record problems and to improve quality following Deming's Plan-Do-Check-Act
13.	Value Stream Map exists, and bottlenecks are clearly identified and prioritized
14.	No pits, no platforms
15.	No multiple-person processes
16.	Everything must be less than 5 feet (1.5 meters) in height
17.	Move as many inspection processes upstream from final test as possible
18.	Utilize as many sub-assembly lines as required to reduce the length of main line
19.	Create Andon system for problem identification and containment
20.	Management and Operations must sustain these principles

Source: Shingijutsu.

do, but we'll get creative and strive to achieve all of them in every process element we design.

"So, let's review," said Ken, as he began to take the team through the principles not found in the "Do's and Don'ts of Manufacturing."

Sylvia raised her hand.

"Yes, Sylvia," said Ken.
"Do you really think we can meet all these requirements?" inquired Sylvia.
Ken replied, "I do. I have seen many teams before you meet most of these desirable conditions, and I have great faith in all of you. Trust me and trust the process; we'll help you with methods and thinking that will allow us to achieve this."

Ken knew his next point would result in some pushback.

"We need processes that don't require forklifts, cranes, and hoists," he began.

The words were barely out of his mouth when George objected that this would create a very unsafe environment.

Ken took the opportunity to reassure the team that, in fact, quite the opposite was true—if they did it right.

"Through the '7-Ways' thinking, you can develop the means to eliminate the need for such devices and make for a safer work environment. There are several issues with cranes. Typically, there is wait time related to them because they are often shared between operators. Then there's the time to connect, move, and disconnect, creating excessive time to move the part. There can be downtime on this equipment, requiring significant maintenance at times. They must be inspected periodically. And contrary to common belief, they pose a huge safety risk because there's always the possibility that the part can fall."

Ken thought George would really go off the deep end when he talked about not allowing for any pits or fixed platforms. These were "monuments" that were not easily moved, and limited flexibility and continuous improvement, and once again created safety hazards. George seemed satisfied with Ken's explanation.

Ken continued, "We need to mistake-proof processes as much as we can, so employees can't do the task incorrectly. Not everything can be 100% mistake proofed, but the closer we can get to this, the more we can reduce the risk of quality defects.

"Next, we need to ensure that operators have no reason to leave their station. We must make sure they have everything they need. We'll have people in roles to support the operators on the lines by ensuring that they have a continual supply of materials. You need to think of the product flow as a river. Water never flows back upstream. We need to ensure that parts and assemblies continue to flow forward, downstream through the value streams.

"Of course, we'll develop standard work early in the mock-up and simulation process, and we'll keep it updated throughout the further simulations. When we launch into production, all of the standard work will be complete, up-to-date, and in place. All direct labor roles will have standard work; this will include all material flow processes. We'll talk about this later, but we'll be developing a plan for every part (PFEP) and simulating and improving those flows as well. We won't wait until the line has been set up to address these items. We need to develop comprehensive training programs where every operator is certified to do many different standard work elements within the factory. We ultimately would like to install a moving line. It's the best way to act as a consistent pacemaker through the process. But we may start with a pulse line—where the line moves together but at set intervals instead of continuously—until we work out all of the kinks that make continuous flow impossible.

"We need to work hard to treat the equipment right and to develop effective total productive maintenance (TPM) practices and processes with the goal to eliminate all unplanned equipment downtime. We'll do this up-front as well with 3P. Of course, we'll demonstrate excellent 5S throughout the process and in the design of the production line. You've done well in this area in the existing lines, but there are several levels of improvement to go. For example, our goal is to eliminate the need for trash containers on the shop floor. Parts should come out to the lines without packaging to be removed and discarded.

"We'll also include a quality-at-the-source program in our new production process to ensure that there's a process to detect defects as close to the source of creation as possible. We'll put a 'firewall' in place so defects won't flow downstream, until they can be eliminated completely. It won't be a real wall, but rather a policy that the operators will be expected to follow when quality problems are uncovered. The firewall will be part of a comprehensive 'Andon' system, in which operators can easily alert a 'first responder' team when they encounter an abnormality such as a quality problem or something else that's preventing the line from moving at the desired frequency, or 'takt time.' An escalation process will be developed, whereby if a 'first responder' team can't quickly resolve the problem, the issue is escalated continuously in a defined manner as time progresses until a solution is determined.

"Throughout the process, time is very important. Keeping times as short as possible is highly valuable to the customer and the shareholder. One thing we need to do is to move as much work as possible off the main flow line and complete it on feeder lines or in work cells. Major sub-assemblies will be built in these work cells which would then be installed in the final assembly on the main line. It's a form of 'plug-and-play' process, with fewer stations on the main line as a result. Given that time through the process is dependent on the number of stations and the number of units in process, this will result in less 'flow time' through the process.

"Team, as you can see, there are many principles we need to work to adhere to. The closer we are to these, the closer we'll be to world-class manufacturing levels out of the gate. I know many of you are a little skeptical. And I recognize it's a lot to take in. We'll be learning as we go, so please don't stress about it.

"There's another key lesson that we need to stay true to. This will be one of our bigger mental challenges, because most human beings have a propensity to always want to jump to the solution when they are asked to solve a problem. And believe me, we'll have many problems to solve throughout this effort. We don't want to jump to solutions. Instead, we want to follow a process that enables us to get to the best solution. So we really must be disciplined in our approach."

Ken then launched into yet another new lesson for the team:

DON'T JUMP TO THE SOLUTION...TRUST THE PROCESS

It's human nature to think that you have the answer early on when confronted with a problem. There are so many "A"-type personalities in business, and in truth some really smart people as well—people who have made their careers with their gut, on just knowing what to do. In many cases, they may indeed have good solutions but they most likely don't always have the best solution early on. Following the 3P way will get us to the best total solution we can come up with. But there is a process, and we must follow the process; we must trust that it will get us to the answer. We must resist, over and over again, jumping to a solution.

Ken spent the next hour taking the team through a high-level view of all elements of the 3P process He introduced them to the "Process-at-a-Glance" and "7-Ways" methods. He would review these in much greater detail prior to taking them through the "real-world" exercises at the appropriate time.

Brenda Lewis, the product design engineer assigned to the team, had been sitting silently toward the back of the room throughout the training. She said,

"Everything I heard you cover today had to do with manufacturing—one-piece flow, equipment design, and so on. While I find it all very interesting, I have to ask 'why am I here?'"

Ken smiled. "You, Brenda, are our 'ace in the hole.' You're an engineer, right?"

"Yes," Brenda responded.

"And engineers are good problem solvers."

"I like to think so," Brenda said modestly.

"As I said before, Brenda, we'll have many problems to solve. For a particular problem, you may very well have a solution that exists in the product design itself. In other words, a simple change in the design may help us overcome a difficult obstacle. Do you now see the important role that you can play on the team?"

Brenda responded, "I think so," still a little unclear but willing to give it a try.

Ken cleared his throat, making sure that everyone would hear what he was about to say:

"The ideal 3P project would involve the concurrent design of the production process and the product itself. It's true that a design for the Trail Gripper product already exists, so we won't be starting from scratch. Nonetheless, we have a great opportunity to significantly improve it. We'll learn 'design for manufacturability' concepts that, if incorporated in the product design, can simplify the production process, improve quality, and reduce cost. The '7-Ways' technique that we reviewed can even be applied to aspects of the product in order to identify the best solution to a design problem. Brenda will be our liaison to Product Design Engineering. I expect her to be quite busy throughout the 3P process, as well as several of her fellow engineers."

Brenda smiled at the sound of that.

❋   ❋   ❋

Dave was sitting in the family room with his wife, about to have a steaming after-dinner coffee, when their three-year-old Golden Retriever came running up looking for some attention. Julie asked,

"How was work today?"

Dave said, "Not bad. We had our first class on this new 3P process."

Julie interrupted him. "I heard you talk about this before. Why are you saying this is new?"

Dave responded, "Well, I'd always thought I was doing 3P. But whatever I was doing before wasn't really 3P."

"Is that a bad thing?" his wife asked.

"No, not at all. I'm really excited about what I'm learning and what we're about to do."

Dave pulled his laptop over to him, opened the lid, and logged on. As his computer came up, he heard a couple of beeps and saw that Pete had sent an e-mail:

"Hey. Dave, didn't get a chance to talk to you after the training class today. I'm even more convinced that we're doing the right thing, but for the life of me, I can't imagine how we'll do all of these things Ken taught us today. When we do, this is going to be an out-of-the-park home run. No need to respond. Enjoy your evening at home. We can talk in the morning. We're about to go for a ride!"

❈ ❈ ❈

Chapter 6

The Right Start

Steve Sawyer, Enterride's vice chair, took a last glance at the menu, trying to decide between a chicken salad wrap or the big juicy cheeseburger, lately conscious about his weight after one of his brothers had a mild heart attack a month ago. He closed the menu. He'd order after his boss, Frank Kent, did. The two of them had slipped out to have lunch together between meetings. Frank ordered the cheeseburger platter, and that clinched it for Steve; he caved in and opted for the same.

"What's going on at Trail Rider with Pete Grant?"

Steve had been reaching for his water glass. He looked up and said,

"They're starting a 3P process to bring the new Trail Gripper product on line. Pete went over it with me last week; I don't know much about it, but it sure sounded good. He described it as a vertical start-up, launching on time at target quality and target cost."

Frank looked at him more closely, "3P, you said?"

"That's what they call it."

"That's interesting. I'm on the Board of St. Lucia's Hospital Group outside Austin, Texas. We had a board meeting just last month, and Joan Jarrett, their CEO, talked about using 3P to redesign existing operations and to design entire new facilities for new services they're now offering. They've been at it for a couple of years now. Joan swears it's the best process she's seen in her career. They've reduced capital costs 25% this year alone and have gotten rave reviews from hospital employees and patients. In fact, they won several major awards last quarter. Based on what she said, Trail Rider might really be onto something—the breakthrough we've been looking for at the Memphis operation. Let me know if there's anything I can do to help."

Steve made a mental note to go to St. Lucia's to learn more about the process. Pete had told him that 3P is one of the best design processes out there. If they were using it in a hospital, it had to be applicable to many things.

"Frank, do you think I could get the Trail Gripper team down to Austin to benchmark what they're doing down there?"

"I don't see why not. Send me a reminder, and I'll call Joan. I'm sure she'll be fine with it."

❋ ❋ ❋

"Welcome to the future," Ken Saguchi smiled from the front of the new Trail Gripper team room. Pete Grant had freed up the full team for the 3P project, and they were all in the room for their first day of activities. Pete had committed to spend the entire first week with the team; the part-timers would cycle in as needed.

"This is going to be our home for the next year."

The 3P team was in a rectangular room near the back of the building; it measured 25 × 50 feet and had two doors, one entering from the far side of the warehouse and the other double-door opening out into a 25,000-square-foot area. Nearby, a construction team was starting to pour a slab for the new expansion. The team room was still filled with some old filing cabinets and a variety of tables and chairs, and two of the walls were littered with the stories of the past 20 years of the plant, or at least it seemed like it. There were numerous old memos, what looked like out-of-date work instructions for long-gone processes, and a calendar from ten years ago. These two very long walls would be perfect for many of the visual displays they'd be creating and posting for continual reference throughout the project. Ken continued,

"First things first. There's nothing that we'll be doing here that won't perfectly emulate the essence of a true Lean environment. This is a process, the same as any other process, and all the principles and Lean practices will apply. The first thing we'll do is to put our new home in order with a big 5S push. We're going to *sort* out everything that we don't need, and get what we do need. We're going to create a place for everything, '*set* it in order' as we say. Then we're going to make this place *shine*. See those cobwebs along the ceiling? Lord only knows the last time this place was cleaned. We're then going to build *standards* so that where everything goes and how it will be used are defined. Most important, once we do all this work, each and every one of us is going to *sustain* it. Everything we need will have a place, and throughout the project everything will be in its place if we're not using it. I'm going to be a task disciplinarian on this, but we all need to be vigilant. Are we ready to get going?"

Mary raised her hand.

"What about *safety*, the 6th S? I once read an article that listed six."

"Great point, Mary. The original Toyota Production System or TPS focuses on the 5S's. I know that some companies have embraced *safety* as another S. Safety is clearly important. In fact, one of my own senseis at Toyota taught me that safety and quality permeate all aspects of TPS, all the time, anywhere, with everyone. In one form or fashion, every element of TPS embodies the principles of safety and quality, and 5S is no exception. My early teachings convinced me that we don't need to add another S to make it clear that safety is at the root of every step of our 5S work. But I'm glad you raised that, because each and every one of us needs to make safety paramount during this project. We'll start every day with an awareness message, and we'll all take a pledge to think and act safely in all we do. We'll look out for the well-being of our teammates, and we'll build safety into every part of the process we're developing for all of our colleagues who will be making the Trail Gripper after launch.

"About half of the team will work on this room. We want to clear all the walls of everything that's up there and wash them down. We want to clear out all of the furniture and fixtures we don't need. George has found a spot to put these items, but he's asked that we red-tag each of them with the date and store them neatly. He'll show us where he wants everything, and he'll discard them in a year if nobody else needs them. We'll have three fold-up tables in the middle of the room, with about 18 stacking chairs, which I want stacked against that wall in three rows, six high. We'll be standing for much of the next two days. We'll also want three more folding tables close to that side wall, where we can sit down to do some table work when needed. We have ten easels in the other room. Let's bring six of them in here and set them along that long main wall over there. We'll leave the other four easels out there. They'll come in handy during our simulation work.

"The rest of the team will work with me on our big work area outside of our team room. We want to create an area for supplies, because we'll use a lot of different materials during the project. We want them to be stored neatly and be easy to find, and we'll have a simple replenishment process for each of them. George brought in three shelves. We have two cabinets and some floor space for larger items. For the past few weeks, a few guys from the materials team have been dropping off all kinds of supplies that Dave and I requisitioned. Between the two of us, we'll explain what they're for, and where we should store them and how.

"We have different sizes of paper, rolls of Kraft paper, Post-It® notes, index cards, lots of pencils, colored markers, glue, wood, plastic dowels, foam, cardboard, steel, pipe banding, wire, Creform® modular material handling and storage supplies, and a bunch of other things."

Sylvia remarked, "I can't even imagine what we're going to do with all of those things."

Ken said, "Trust me, Sylvia; we'll all understand before too long. We're also going to set up a workshop. George has helped us pull together three work-benches, two table saws, vises, several shop vacuums, a pipe cutter, drill press, and all the hand and machine tools that we'll need. We'll need to take some time to build shadow boards for all the tools so that they're easy to find and easy to use."

※　　※　　※

Ken split everyone into teams, gave them a two-hour target, asked them to divvy out the tasks among themselves, and sent them out to work. Within five minutes, both areas became a beehive of enthusiastic activity. The team was brimming with positive energy.

The team room was set up about 15 minutes ahead of target. Several of the team members were used to working in a very clean office environment and were determined that their new work areas would be no different. The team had taken *shine* to the next level with buckets, mops, and a variety of cleaners. The whole team room smelled like a hospital ward, and it sparkled. The storage area and workshop were almost as clean and neat, and were certainly a better example of 5S than any other area in their big factory. Ken was pleased. The team members were pleased, and they were told to take a break and reassemble in the team room in 20 minutes. Pete suddenly remembered what it was like to get his hands dirty and had a newfound respect for all the operators in the shop. He felt like going back to his office for a nap, but wasn't about to be the only one not back in 20 minutes to pick up where they left off.

Once the team was reassembled, Ken said,

> "Okay, team. I'm going to ask you to do perhaps the most important part of our mission. You might find this hokey, but I need each of you to take a few minutes right now at the start of our work together and make a deep commitment to yourself, your teammates, our project, and your business. Commit to how you will think and behave throughout the project. Pete has some expectations that he's to ask you to commit to. If you truly can follow them throughout the project, then I'm going to ask you to sign a contract to your peers that represents your commitment. If you don't think you can abide by these expectations, we want you to do what's right for your team and say so. Pete has committed to replace anyone on the team who can't agree to the rules and guidelines and wants to be removed from the team. We're very serious about this work, and we need rules; otherwise our objectives are at risk. If you can't agree, please do the right thing and step off the team. There won't be any consequences for you, and you'll be much happier than you would if you were held to principles you can't support."

Ken handed everyone a copy of the contract (Exhibit 6.1) and asked them to read through each item, reflect on them, and decide if they could make that commitment.

3P **Production Preparation Process**
Building Quality into Our Production System

Contract

To give YOU permission to be creative

In order to solve the problems we have facing us and to produce the results we desire, I must think and act creatively. I promise to do the following to produce creative, productive results:

- *I will try other people's ideas, not judge them (good or bad).*
- *I will make every effort to overcome the hurdles in my own thinking in order to generate useful ideas. I will look to nature for fresh perspectives and proven solutions.*
- *I will work within the group to accomplish and deliver what was promised.*
- *I will not be constrained by my job title, education or position, nor will I constrain others. Instead, I will encourage others to produce ideas which are different, contrary, and better than mine.*
- *I will express my ideas openly and not fear laughter, judgment or failure.*
- *I will return to my 12 year old mindset and will use this style of thinking to generate useful ideas.*
- *I will focus on creating flow in all seven elements.*
- *I will eliminate waste ... even if it is carefully disguised as work.*
- *I will never settle for the first idea but will instead choose the best idea from at least seven.*
- *I will quickly and crudely test ideas on the shop floor and on the actual product.*
- *I will not substitute money for brains and hard work.*

Signature:_____

Date:_____

Coach:_____

Date:_____

Exhibit 6.1   Team contract. (*Source*: Shingijutsu.)

After a few minutes, he read through each item with the team.

1. *I will try other people's ideas, not judge them (good or bad).*
   "It's important that each of us commit to be open to all ideas that come forward."
2. *I will make every effort to overcome the hurdles in my own thinking in order to generate useful ideas. I will look to nature for fresh perspectives and proven solutions.*
   "All of us have subconscious barriers to new thoughts. We all have our own biases. We need to work hard to set them aside. We also can find the purest solutions to problems throughout nature and we need to seek these solutions."
3. *I will work within the group to accomplish goals and deliver what was promised.*
   "This is important for every project. We will set targets, measure progress frequently, and stay on schedule."
4. *I will not be constrained by my job title, education, or position, nor will I constrain others. Instead, I will encourage others to produce ideas that are different, contrary, and better than mine.*
   "We need to set aside any semblance of rank while working on this team. We're all equals. We also need to recognize there will always be a best way, it won't always be mine, and we need to be okay with that."
5. *I will express my ideas openly, and not fear laughter, judgment, or failure.*
   "Not everything will work. Not everything will sound like a reasonable idea. If we're ultimately to be successful, each of us needs to put every idea we can come up with on the table, whether it'll work or not, or whether others might think it's crazy."

6. *I will return to my 12-year-old mind-set and will use this style of thinking to generate useful ideas.*

   "I'm reminding all of you of this key point from our training. We need you to think like a 12-year-old all the time and think about all possibilities. It's important to our success if we're to come up with breakthrough ideas. There will be times that you will revert to your adult ways. Therefore, you need to constantly remind yourself of this need. I'll be reminding you also."

7. *I will focus on creating flow in all seven elements.*

   "I'll review the seven elements of flow in detail later. Generally speaking, our goal throughout this project will be to ensure that all information, all materials, and all production work-in-process flows continuously through the process at the pace or takt time that we establish. We need to eliminate all areas where flow stops."

8. *I will eliminate waste, even if it is carefully disguised as useful work.*

   "Remember, we have a chance to not put waste in right from the start. As we kaizen, simulate, and observe over and over throughout the project, we need to see waste and eliminate it."

9. *I will never settle for the first idea, but will instead choose the best idea from at least seven.*

   "I'm going to introduce you in more detail to 7-Ways, but basically we'll work as teams to generate at least seven ideas to solve a problem or a need. We need to be open to picking the best one, not necessarily the one you and your team came up with."

10. *I will quickly and crudely test ideas on the shop floor and on the actual product.*

    "When we come up with the best solution, we need to commit to mock up and test out the idea in a simulated work environment. This goes well beyond the 'air guitar' practice that Dave has taught you. Eventually, we'll do the same thing with the actual components and assemblies of the actual product."

11. *I will not substitute money for brains and hard work.*

    "We need to follow the principle that we have no money. If we get creative and work hard, there are always ways to solve problems for much less money than we might have traditionally solved them."

Once they went through the 11 commitments, Ken asked the team to sign their personal contracts. Each of them did. He then asked Gina Nelson to neatly tape them in a group on a side section of the wall, telling the team that these contracts would be posted visually for the duration of the project as a continual reminder to each of them of the commitment they'd made (see Exhibit 6.2).

"We still have a bit of time before lunch, and there are some things we can get done beforehand. First, do you all remember the 'Do's and Don'ts of Manufacturing'? I hope so, because we all need to think about them continuously in every aspect of the process we're about to design. We're going

Exhibit 6.2   Posted contracts.

to post these on the wall right beside the contracts. They'll act as a visual reminder, throughout the project, of what we'll strive to do and what we'll work hard *not* to do. John, can you please put those up on the wall?"

Ken asked Mike Young to grab an easel and bring it over to the team. He quickly created a table of sorts and asked Mike to write all of the team names in the left-hand column and post it on the wall.

"We're going to keep our team list up on the wall, and we're going to track attendance every day. It'll be a visual way for folks to easily see who's in on that day, as team members may be working in different parts of the plant at different times during the project. It also reminds everyone that attendance is important (Table 6.1). The good news is, for Day 1, we have perfect attendance."

And he picked up a different marker and put a green "X" beside each name under the first-day column.

Bill Stark, the material manager, asked, "Is there a reason you used green?"
"I'm really glad you asked that, Bill. You're the perfect setup man. On our charts, when something is normal, we're going to mark it in green. If it's abnormal, or a 'miss,' we'll mark it in red. We want to reserve green and red for a differentiation of normal and abnormal. All other colors can be used for any other purpose."
Bill asked, "Even in here, we'll stick to the red and green?"
"Yes, Bill, I'd advise all of you to use this same practice with all of your processes going forward," said Ken.
"It seems like kind of a simple chart," Sylvia observed.
"Exactly," remarked Ken. "We're not looking for slick and fancy trackers for the wall. We want them to be simple and easy to use, maintain, and update. And they should be understandable to everyone.
"We still have 30 minutes until lunch. Let's start developing our schedule. We can complete it after lunch, and certainly by the end of the day."

Table 6.1   Attendance Tracker

Project Trail Gripper Attendance Tracker							Week: 1
Team Member	4/3	4/4	4/5	4/6	4/7	4/8	4/9
George Hall	X						
Mike Young	X						
Gina Nelson	X						
John Lee	X						
Bill Stark	X						
Sylvia Bennett	X						
Lou Marks	X						
•	•						
•	•						

George started to leave the room, but before he got all the way out the door, Ken asked him what he was doing. George explained that he was going to get his laptop to build the schedule model. Ken softly took the opportunity to remind him and the team that while they would definitely be using computers for some things, much of what they did would be done manually and put up on the wall, large as life, to make it easier to review and update. Everyone nodded. George, the advanced manufacturing engineering leader, thought they were going back to the 18th century.

Ken had a few people unroll about 20 feet of Kraft paper and start taping it to one of the long walls, while others grabbed some Post-It® notes, index cards, string, and black markers. While they were doing this, Ken described the process they would be going through after lunch. He told them they were going to start with the target launch date, walk backward to this week, and map the flow of the work by month, with all the interconnections also mapped out. They would also develop a detailed schedule that they would divide into three-month intervals, which would allow them to track progress daily.

Lunchtime came and Ken called a break. Pete wandered down to his office with Dave before they joined the team in the lunchroom.

"So what do you think?" Dave asked Pete.

"Well, it's a lot more detailed than I thought it was going to be, but the approach is straightforward enough. I like it."

"I thought George was going to have a stroke with all the paper up on the walls."

Pete laughed. "I know. George is the master of the spreadsheet. His team says they think he has his computer surgically implanted in his side."

Dave commented that the team seemed to be pretty engaged all morning.

"Yeah, they sure were. I've never seen this group work so hard as when they were doing the 5S prep. Everyone was getting along and having a lot of fun. I think this is going to be great. Let's grab some lunch."

❋   ❋   ❋

After lunch, the group gathered back in the team room. Ken had the team develop the schedule together, because he wanted everyone to understand it, as well as to have collective ownership of it. He asked the team to identify the best scribe. Everyone's heads turned toward Linda Campbell, the supplier quality engineer. Betty King, one of the commodity leaders, affirmed that Linda's handwriting was perfect. They had their scribe.

"Let's evenly spread out the top row in monthly increments for 12 months, and we'll put a big star up there on top of February 27, which is our target launch date. That's our launch date, and we won't miss it. We'll add the major stages of the process along the left-hand column. We'll break down the process into ten stages, so why don't a few of you grab a long straight-edge and mark ten discrete, equal-height rows and column markers between each month as guidelines?

"So Linda, start writing down the following descriptions of the tasks: Target Preparation, Benchtop Mock-up, Scale Mock-up, Information Flow, Material Flow, Full Mock-up, Equipment Design, Equipment Build, Equipment Simulation, and Launch Prep."

Under Ken's guidance, the team went on to build a simple and visual project schedule for each task. Sub-tasks were identified, along with a timeline for each. The sub-tasks were written on Post-It® notes, so they could move them around as other parallel elements of the schedule were added. The team used string to display the interconnections between tasks and clearly showed linkages between interdependent tasks, and key times during the project when information and material flows interacted with the various mock-up stages. By the end of the first afternoon, they had crafted a full schedule (Exhibit 6.3) and headed home for the night.

The next day, the team once again gathered in the team room. They were all there, all on time, and all excited to see what was next. Ken didn't waste any time doing attendance checks and launched right into Day 2.

"We're fortunate in some ways—not so fortunate in others—that we're moving an existing process into a new empty footprint. This allows us to do comparisons, to have access to existing data on current processes, and to look for areas to improve and eliminate waste. However, there are some existing paradigms, and Enterride may not be so open to not using some of the existing equipment from the old Trail Gripper factory. Dave took a team

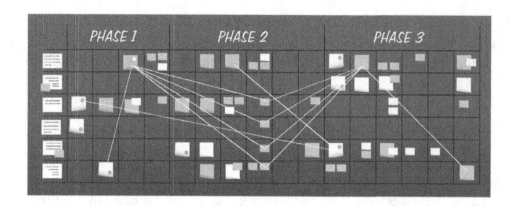

Exhibit 6.3   Current Trail Gripper project schedule.

out to Trail Gripper and spent a week there performing a review as part of the acquisition process. They have a highly automated process, big monuments for equipment, a lot of conveyors, and minimal flexibility. We're not going to want to use a lot of the equipment they have, but there are some fixtures and tools that we may use. The good news here is that most of their equipment is depreciated, so the finance folks won't be as tough on re-using existing equipment, plus corporate has identified a buyer and they want to run the factory as-is, producing a low-end version of the Trail Gripper. Enterride is going to maintain a 40% share in this joint venture, so the low-end Trail Gripper will still be part of the corporate marketing plan, but we'll build all the medium- and high-end units here. I want us to develop the process with an open mind and not be predisposed to existing equipment, fixtures, and tools. We may choose to use some of the items from the old Gripper factory, as long as the joint venture doesn't need them.

"So, last night, I had John post some of the data that they accumulated at Trail Gripper. I want to start there and review it as a team. I understand that most of you were able to go there for a couple of days over the past few weeks. That's great. It'll help a lot that you were able to observe much of the current process, but not be involved long enough to get locked into any parts of it. Let's start with a list of the high-level stations they have and a comparison of process times for each step, by model number. We'll use this as a baseline, but work to improve the process times for each as we work to reduce waste. Let's look at the process time chart (Table 6.2). What do you notice?

"There's a big spread in process times for many of the steps, particularly 3 and 11. I'm not sure how we'll be able to flow them down a single line together and maintain takt time. I guess we'll need multiple lines," said Dave.

"George, that's a good observation. There are indeed a number of areas where there's a lot of variation, and we'll need to take that into account. However, we do need to be careful not to jump to solutions. We'll follow the process and get to the best answer for each of these challenges.

Table 6.2 Current Trail Gripper Process Times, by Station, by Model

Station #	Model A	Model B	Model C
1	4.0	4.1	4.1
2	4.8	4.7	4.7
3	5.6	6.2	5.7
4	3.5	3.6	3.5
5	4.3	4.4	4.3
•	•	•	•
•	•	•	•

"Let's look at the barchart beside the process chart (Exhibit 6.4). This is a way to lay out the same data in a more visual format that can better show balance and variation of process times against takt time. Let me explain it for those of you who may not have seen something like this before. The left axis represents time in minutes. The solid black line across the top is the current takt time at the Trail Gripper plant. This is their current production rate. The dashed black line below is the takt time that we'll be designing our line for. It's 10% faster than the current takt time. We'll talk about that in more detail this afternoon or tomorrow. All the stations on their main line today are shown along the bottom. There are two process times displayed for each station, using lines or bars. One is the minimum process time for the fastest model at that station, and the other is for the maximum process time for the slowest model at that station. Does anything stand out for you?"
Bill Stark chimed in, "That doesn't look like a good balance on these lines. The process times vary a lot from station to station."

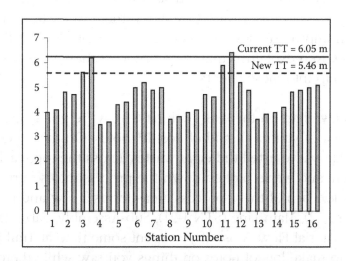

Exhibit 6.4 Barchart of current Trail Gripper process time and takt time.

"Great observation, Bill. We'll be referring back to this chart often in our early work as we properly balance our line design to meet our target takt time. At the appropriate time, I'll introduce another graph called a Yamazumi Chart."

"A Yama-what?" chuckled Sylvia.

"A Yamazumi Chart. Some people call it an Operator Balance Chart. You'll see what it is later."

"Let's try another exercise. How many people here know what the seven wastes of Lean are?"

Everyone raised their hands, except for moonshiners Norm Wilson and Bill Cook, and Betty King, the metals commodity leader.

"Okay, so some of us haven't heard of all of them. I think it's worth a quick review, because these are so important to the work we'll be doing. Our goal will be to eliminate as many of these wastes as possible from the processes we're designing. You remember our conversation around 6S versus the original 5S? Similarly, some people have added an eighth waste to the seven-waste model. For the sake of our project here, I would like us to refer throughout to the full eight wastes. I've found the addition of the eighth one to be very helpful. There's a simple memory jogger you can use for the eight wastes; in fact, I wrote this out on a flipchart, and we'll post these on the wall as well (Exhibit 6.5). Just remember the acronym 'DOWNTIME.' Our goal is to make our processes flow continuously and never be down."

Ken went on to again explain the eight wastes to the team, teaching some and reminding others and elaborating on the nonutilized talent waste—the eighth waste that some have added to the traditional seven. He explained to the team that an organization often fails to utilize its people's full abilities, particularly their mental and creative abilities. In other words, a company doesn't get the benefit of all of their observations and ideas, the value of which is enormous, and not having access to it is a waste. He asked them to recall how often they were involved in improvement efforts before Pete Grant's arrival. People nodded in understanding. Ken explained that the 3P process they were going through was one way to tap into these talents.

"Okay, on a new chart on the wall (Exhibit 6.6), we're going to go through an exercise. Each of you has a pad of Post-It® notes. On the sheet on the wall you'll see that we've listed the eight wastes across the top, each with its own column, and we've broken down the process into several areas on the left-hand side. Each of these process areas forms a row and is listed as Main Line (STN 22-42), Main Line (STN 1-21), Sub-Assembly, Fab, Information Flow, and Material Flow. Most of you spent some time at Trail Gripper, and I asked you to make lots of notes on things you saw while there, both opportunities for improvement and non-value-added wastes. We're going to take

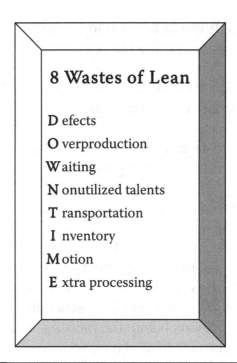

Exhibit 6.5   Eight wastes of Lean.

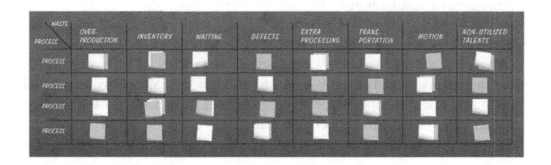

Exhibit 6.6   Wastes in the current Trail Gripper process.

30 minutes. I want you to identify all the wastes that you saw there. Write them on the Post-It® notes and then pile each note under the appropriate waste and process area. Look at what others are writing; it may trigger other ideas of wastes for you. We'll keep this up on the wall as well and refer back to it regularly. We want to ensure that these aren't in our new process design. If there are no other questions, let's begin."

After 30 minutes, the team had accumulated about 150 examples of waste in the current Trail Gripper process. Ken went through each of them quickly with everyone for another 45 minutes, asking for further clarification of any that were confusing. He then taped them together in their appropriate sections. One of the most significant wastes was a yield problem at Engine Start/Test. Six percent of all units produced on the line failed this test and required a significant effort to repair. Nobody knew at this time, but this list would prove particularly valuable in the upcoming months. It would be used to trigger memories of how to make the new process much better than the old one.

Ken said, "That's enough for now. Let's break for lunch." A lot was covered in the morning, and people needed a break.

After lunch, the team re-assembled in the team room. Almost immediately, Blake Derry, one of Enterride's marketing managers, came walking in.

"Hey, Blake. Thanks for joining us."

"No problem; glad to be here."

"Team, I asked Blake to come in this afternoon to walk us through our next exercise, which we'll try to do in the next 45 minutes. We want to understand the markets for this product transfer over the next few years, particularly the estimated volumes and ramp-up schedule. I asked Blake to draw out his ramp-up estimations before the meeting, and I have it on the wall right over here (Exhibit 6.7). This is a two-way conversation. It's our chance to challenge some of the assumptions, as well as to get a full understanding of the things we don't understand about them. Blake, why don't you walk us through your assumptions?"

"Sure. We've been in touch with many of Trail Gripper's key customers and because this is mostly just a product transfer, we won't see the typical ramp-up that would happen with a new product launch. We expect most customers to stay with us as we transfer the product. Most have expressed optimism, and we hope to see some growth in market share. We're expecting an initial 30% volume decline because we've agreed to have another company produce the low-end product line as part of a joint venture.

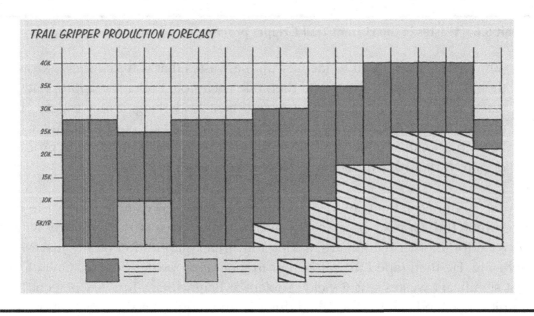

Exhibit 6.7   Trail Gripper volume projections.

However, we're adding a couple of options that we think will be popular with our customers. We expect the options to offset that 30% decline over the first nine months. We also estimate a 10% volume gain from increased market share, due to higher quality and delivery performance out of this plant. Personally, I want to thank all of you for your work the past year. I've been with headquarters in sales and marketing on the Trail Rider product line for the past five years. If we'd had this conversation a few years ago, I'm sure we would have avoided the near-total revolt from our customer base and the substantial drop in market share that we experienced. You've clearly made big improvements in the past year. Our customers recognize it, and I'm hearing positive comments almost every day.

"You can see in our initial estimates that we'll start out of the gate at a rate of 28,500 a year, which would be equivalent to 2,375 a month, or 548 a week. We're projecting that in about nine months we'll have gained back all 30% of the initial volume decline, to take us to about 37,050 a year. This would be about 3,088 a month, or 713 a week. You can see in our assumptions that we anticipate a straight upward slope from launch to the ninth month, but we're really not entirely sure what the ramp curve will look like. We do think it'll be pretty close to this, though."

"How did you arrive at this, and how confident are you that this is how demand will come in?" asked Dave.

"Good question. Of course, there's no way we can predict with complete accuracy what will happen with a new product family, particularly this far from launch. We sent out a task force of five of our best salespeople, and they talked to 20 of our biggest customers about their plans when we launch the product here. In almost all cases, our team was confident that the customers gave us credible information about their intentions, although they themselves can't say with any certainty the volume that they might need. They did provide their best guesses. We have a lot of customers, and we projected what the others might do based on our top 20. We feel good about the figures.

"On top of this, we're confident that will get to a rate of 40,000 units a year by the end of the first year, and this would put demand at around 3,333 a month, or 769 a week. After that, we think we'll level out for about six months; a number of agencies are estimating a temporary saturation. The project was approved based on a 5% average annual growth rate from then on. There's no reason we can't do this."

Blake asked, "What do you think Pete? Will you be able to keep up with these projections?"

Pete chuckled, "You just bring it on, and we'll deliver."

"Pete, I'm confident your team here will be able to do it, but maybe you should go back to your office and let them work it out on their own. You might get in the way."

"Nice Blake, come back any time."

Ken said, "We have our ramp-up plan. What capacity do you think we should plan for, Pete?"

"I'm not sure; maybe we should find out where they think we'll be in three years and make sure we have capacity for that. What do you think, Ken?"

"I'm confident that your facility expansion can accommodate the projected increase in demand for the next ten years at least, with continuous improvement and workplace compression. Normally, we wouldn't recommend planning for any more than the next 12–18 months. We can reduce takt time and develop another production scenario pretty quickly when volume increases after that. I suggest we plan our process here for the 40,000-a-year rate."

"If 12–18 months is the general rule of thumb, then that sounds good to me," Pete said. "Anyone else have any objections?"

"Are we sure we can ramp up quickly after that? I would have thought we should plan for extra capacity while we're doing this project," said Gina Nelson.

Ken responded, "I'd say if we don't need the capacity in the short term, then planning beyond that would be waste."

"I never would have thought about it like that before," said Gina, "but it makes sense. This is not much different than over-producing to schedule."

"Anyone else have any different thoughts?" Pete asked.

A chorus of "no's" quickly echoed through the room.

"That settles it. We'll plan for an annual 40,000 rate. I want us all to walk through this next exercise together. It's very important, and much of what we do will derive from it. I want us to calculate our takt time. John, you can be our scribe while we figure this out."

John walked up to an easel.

"How many per week does 40,000 a year translate to?"

"I wrote it down. Blake said it was 769 a week," Sylvia said.

Bill Stark had a look on his face of a man who had something on his mind. He spoke up,

"That's based on a 52-week year. Pete, I heard we were considering adding a two-week shutdown. That would give us only 50 working weeks a year."

Pete responded, "I remember we had that conversation about a month ago. We discussed it with corporate, and they strongly advised against this. They felt that our relationship with our customers was still shaky and didn't feel that it was wise to disrupt servicing them for two weeks. Even if we were to build up finished goods before the shutdown, it would be impossible to stock all of the possible options that customers might order. It's good you were thinking about this, though."

Ken spoke next: "How many hours do we work a shift?"

"Eight not including lunch."

"How many shifts per day?"

"Two!"

"How much time do we take off per shift?"

Mike Young responded, "Well, we have two 15-minute breaks, a 15-minute meeting and prep time at the start of each shift, and a 15-minute clean-up at the end. That's an hour."

"So we have seven available working hours per shift. At two shifts a day, that gives us 14 available hours per day. John, capture all of these figures on the chart. Let's keep going. We have a five-day workweek, so we have 70 available hours per week. We should convert this to minutes for a takt time calculation; 70 hours is equivalent to 4,200 minutes. John, do you know how we calculate takt time?"

"Sure, we take the available hours and divide it by the demand over that time period."

"So, what will our takt time be, based on this?"

John took out his calculator, then wrote the last equation on the flipchart (see Exhibit 6.8) and told the team that the number worked out to about 5 minutes and 28 seconds.

"That's great! So does everyone see how we got to this number? If not, don't be shy; I want to make sure that every one of us understands this; it's one of the fundamental factors we'll use."

Everyone nodded understanding. Another big lesson learned!

"John, let's put the takt time calculation up on the wall. We'll refer back to it regularly.

"Good work, team. We have another two hours until the end of the day. I suggest that we use that time to finish up any items left undone during our 5S work yesterday. Then we'll call it a day and get a fresh start in the morning. Tomorrow we start planning our flow process."

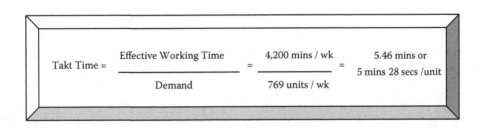

$$\text{Takt Time} = \frac{\text{Effective Working Time}}{\text{Demand}} = \frac{4{,}200 \text{ mins / wk}}{769 \text{ units / wk}} = \frac{5.46 \text{ mins or}}{5 \text{ mins } 28 \text{ secs /unit}}$$

Exhibit 6.8 Takt time calculation.

Chapter 7

A Benchtop View

The team assembled the next morning, eager to see what the day would bring. Ken kicked things off:

> "We're going to start laying out our flow lines; but before we do that, we're going to set down our targets together in writing, and we'll post them on the wall in a prominent place to remind us every day what we've committed to deliver. Keep in mind that we own our targets as a team. Our objective is to meet all of our goals, period. I captured what we originally signed up for and wanted to go through them one by one, and make sure we all still agree. This will be our last chance to debate. If we feel they should be changed for whatever reasons, it's important that none of us be shy and that we speak up. There is nothing wrong with push-back; it's better, and everyone has a voice in this. We may not change any of them, but we certainly should openly discuss where any of us think we should. Sylvia, please help me tape this up to the wall." (See Table 7.1.)
>
> "You can see that we have the production start-up date of February 27. This is a 'can't miss' date. We'll launch on time, which means that we'll be at full production for current demand, starting on that date. Any objections?"

None were voiced.

> "Capital expenditure for the project was originally estimated by the Advanced Manufacturing Engineering group to be $42 million. We committed to reduce that by 20%, or $8.4 million, to a target of $33.6 million. Any objections?"
>
> George spoke up. "I know what we signed up for, but I did the original estimate. I've done many line relocations like this over my career, and I can't see how we can do it for $8.4 million less than the original estimate. I think we'll be way short of money for everything that we need to do. You said yourself that we won't be using most of the existing equipment."

Table 7.1 Project Trail Gripper Final Objectives

Objective	Target
Project Completion Date	27-Feb
Project CapEx	$33,500,000
Project Expense	$5,120,000
Product Cost	$4,459
Initial Quality (ppm)	500
Initial Delivery (lead time in weeks)	4
Initial Delivery (on time)	96
Inventory Turns	20
Target Volume/Year	40,000

"George, I respect your concerns. I'd feel the same way if I knew we were going to do this the traditional way, but one of the bigger parts of our project is to develop equipment that is far smaller and simpler, with fewer bells and whistles than Trail Gripper currently has. Fewer bells and whistles means lower cost. Plus, some of your moonshiners will be able to make this equipment for even less. In past projects, we typically saved at least 30% on plant and equipment capital costs, and often it was pushing 40% or more. We only committed to 20%."

"I'm still not sure I'm entirely convinced, but I've seen big changes since we started this Lean journey. I don't like spending money any more than the next guy. If it can be done, let's do it."

"Thanks, George."

Ken continued, "Project expenses were originally estimated to be $2.5 million. An additional $2.62 million was included for the proactive planning, physical modeling, and simulation work we expect to do during the project. It's quite an investment of time for the entire Trail Gripper team to do the up-front work that would normally be done after production start-up. Even at $5.12 million, our payback will be well less than a year. Pete felt we could sell it at this and wanted to keep the additional amount in. I think we'll easily meet the $5.12 million figure. Any concerns, Pete?"

"None. If we don't need the money, we won't spend it, but I'm confident we have all we need plus some contingency."

"Okay, a key commitment is product cost. Corporate originally asked us to take the current product cost of Trail Gripper and reduce it by 3%. We committed to an additional 2% on top of that, for a total of 5% or $234 from the current cost of the product. This puts our target cost at $4,459. How much heartburn does that give you?"

George spoke up again. "I'm less concerned about that than I am about the project capital costs. We looked closely at the product when we were over at Trail Gripper and saw a lot of easy opportunities for productivity improvements and opportunities to reduce material cost. I talked to Betty King, and she thinks that we can reduce material costs through common components and negotiations with existing suppliers."

"So, it sounds like we have consensus on this target?" Ken asked.

Everyone nodded.

"After launch date and target cost, our next really important target is initial quality. There are three primary groups that we try to positively impact with Lean: our customers, our employees, and our shareholders. Significantly improving our performance to our customers helps us look after our employees and our shareholders, and quality is one of the elements customers value most highly. We have a target launch quality of 500 parts per million, or ppm. When we first looked at this project, Trail Gripper was claiming a current defect rate of 1,000 ppm."

Lou Marks, the team's quality leader, jumped in to explain.

"We spent a lot of time at Trail Gripper, looking at their quality data, and that was a clear misrepresentation. This number they quoted was from their best month late last year, and it was an anomaly. In fact, when we looked at their data with them, they acknowledged that the reported number was a huge error. Their three-month rolling ppm has actually been running at a much higher rate, about 18,000."

"Really? Do we believe that?" Pete asked.

Lou responded, "You bet! It's a pretty firm number, and a long way from 1,000."

Pete continued, "What do we think? This is a big change in the baseline of 1,000 we used to set our project target at 500 ppm. It would represent a 97% improvement."

Ken responded, "This must be a goal we all believe in, but I can tell you that one of the most important elements of 3P is to mistake-proof the process we develop. We have one of our best product design engineers on the team now. Brenda Lewis will be here in a few minutes. Once she's caught up on the training, I'm confident that we can identify some minor design improvements to positively impact manufacturability and improve quality. Lou, what do you think? You're the quality representative on the team."

"I'm up for it. I can't see how we can sign up for more than 500 ppm anyway. If we want to increase market share for this product, we have to make big improvements to do it. Besides, we've done pretty well with our own products over the past two years. I recommend that we stick with it, despite the new discovery."

Dave also agreed. Pete knew from past experience that the 500-ppm goal was not just possible, but necessary. Still, it was good to hear the team commit to it.

"Next up, we signed on for delivery targets of four weeks lead time and 96% on time. Any problems with that? It's another performance area that our customers place a high value on."

Now it was Inside Sales Manager Mary Long's turn to weigh in:

"We shouldn't sign up to start at anything less. We've gotten our existing products up to 98% on time, with an average three-week lead time. There's no reason we shouldn't push to get close to those from the start."
"Anyone want to disagree with Mary?"
"No, sir," said Dave.
"Okay, team, that's another one locked.
"The old adage that 'cash is king' is as true today as it ever was. The more inventory we need to carry in order to serve our customers well, the more cash we have to tie up and the less we have available to put to positive work for us. I noticed we'd signed up originally for inventory turns of 20. I felt that we were aiming low on this one, and could do better. What do you think?"
Pete jumped in. "Our corporate inventory turns have been fluctuating between eight and nine for most of the past decade. We've been running around 12 ourselves. I think 20 is a good target."
Bill Stark asked, "Why do you think that 20 is low, Ken?"
"I'm not suggesting starting production at 20 turns is bad. Many people would suggest that 20 is pushing into world-class territory. But my experience tells me that as many as 30 turns might be attainable if we do our work well with *PFEP*, or a plan for every part. It's important that we reach consensus around each target, and nobody should feel ashamed of starting with 20 turns of inventory."
"Let's lock it then. Team, any objections?" Pete asked the group, looking around at everyone and seeing a sea of heads shake.
"Finally, we're targeting a 40,000-unit-a-year rate. We discussed this earlier based on the commercial projections and all agreed to it then, so this one is set. Let's take one more look at our targets. It's our mission as a team to meet them, and it's my role as your coach to take you through a 'learning-and-doing' process to get there. I suggest we start to get into more of the fun stuff. Let's start working on the production flows."

Ken and the team put together four folding tables. They taped together easel paper, two sheets wide and six sheets long, and laid them across the table and gathered the whole team around the sides to begin the process of building a two-dimensional (2D) flowchart.

"So here's what we're going to do. We're going to take the Trail Gripper bill of
material (BOM) and highlight all of its major components and assemblies.
We'll leave out hardware and small items for now. Let's get a copy of the
engineering drawings, and where we can easily do it, photocopy and cut out
pictures of as many of these parts and assemblies as possible. If we don't
have a picture, we'll make a sketch on a Post-It® note. Let's split the BOM
into four sections, and we'll split you into four teams, each one taking a section. Brenda's joined us at the right moment. Hi, Brenda!"

"Hi, Ken. Hi, everyone."

"Welcome to Project Trail Gripper," Mary offered warmly.

"Brenda is our team's product design representative. She's arranged access to
the engineering database and drawings. We have a number of laptops on
the side tables and a network printer in the corner for printing drawings.
There's also a photocopier in the corner. Some of the drawings are over on
that side table; we can photocopy them and cut out parts. Let's go and get
ourselves a full set of paper parts."

Ken then split up the teams, gave each of them a section of the BOM, and put
them to work.

The room was soon bustling with activity. Each team divvied up the work,
with some team members finding drawings and printing, some photocopying
and cutting out the parts into small sections, and some sorting them in order.
The hours flew by. Each team member had zeroed in on the mission, and there
was a palpable sense of excitement and anticipation for what was ahead. This
spirit and work ethic would continue through all remaining stages of the project.
Ken kept a watchful eye and provided coaching where needed. One by one, the
teams finished with their section of the work. Ken gave them a well-deserved
break at this point.

After the break, Ken set the team to work by first explaining to them that every
process has a heartbeat to it, what they might call the pacemaker, and that they
first needed to determine the part of the flow process that would be the pacemaker that dictated the rate of flow for the entire system. Ken had added a flowchart on the wall showing the current Trail Gripper production process with all
the main operations in the sequence that they occur. He told the team to reference
it for thoughts and ideas, but not to get locked into its exact structure, as they
wanted to remain open to finding the best option for the pacemaker. The team
debated and eventually settled on the final assembly process and the components
to flow down it and be assembled, one station at a time, into a final product.

The team then went through an exercise to calculate the total final assembly
process cycles, using the current Trail Gripper process as a starting point. Ken
explained that the current process was a reasonable starting point, and that they
would work to reduce the process times along the way and simplify the flow
paths for speed, margin, and cash. The team calculated that there were around
15,840 seconds of total process time to assemble the current Trail Gripper. They

divided the previously calculated takt time of 5.5 minutes, or 330 seconds, into total process time; it came out to 48. Ken explained that, given the current assumptions, they would need 48 people to perform all of final assembly in order to meet the takt time. Ideally, the total work of 15,480 seconds would be divided evenly between the 48 operators in order to balance the final assembly line. Standard work would have to be developed for each person that contained no more than 330 seconds of work content so that each person would be able to perform the tasks required of them within the takt time. At a maximum, if they had one operator per station, they would need 48 stations for a pulse line, or 48 zones for a paced line.

Sylvia asked for an explanation of the difference between *pulsed* and *paced* lines. Ken explained to the whole team that a paced line was a continuous moving line like you might see in an automotive plant, where the line moved at the pace of takt time and the work-in-process pieces moved along it at the same pace. A pulse line was a line that had individual stations, where work-in-process was placed at each station and at the takt time, and/or when all the standard work was complete, each work-in-process piece advanced forward one station. They could move on a cart, or a conveyor, or air bearings, or be passed by hand, or by any one of hundreds of methods. He explained that although there were pros and cons to both, if the infrastructure support for the process was strong, the paced line was typically better. But they didn't need to decide that yet; they'd get to it in a week or so.

Ken explained that based on the current work content for the final assembly of the Trail Gripper, they would need a maximum of 48 stations. Naturally, each station would have one unit at some stage of assembly to work on. These units are called Standard Work-in-Process (SWIP). Additional stations may be required if the team felt the need to have partially assembled units in queue at particular points in the line. These may be needed if there is some expected variation at a particular station that may temporarily throw off the balance of the line. Additional units of SWIP should be used only after all attempts have been made to address the cause of the variation, and those attempts were not satisfactory.

A goal of their design would be to make the total production lead time through the longest path of the manufacturing process as short as optimally possible. Seeing some puzzled looks, he took the team over to the whiteboard and explained to them why they wanted to reduce production lead time as much as possible. In general, the faster they could move a product from the schedule point at the beginning of the process to the ship point to the customer at the end, the faster and more flexibly they could respond to customer needs. Other benefits were that they'd need to carry less work-in-process inventory, meaning that less storage space would be required. They could wait longer to launch the build process, which would allow more time to receive material. Further, they could reduce the impact of engineering changes or customer order changes. With less work-in-process, any engineering or customer change to the product would have less of an impact. The result would be less rework, lower cost, and minimal disruption to flow when such changes arose. He explained to them some ideas and tricks to do just that:

REDUCING PRODUCTION LEAD TIME IS KING

A simple way to calculate production lead time (also called total product cycle time) is to multiply takt time by SWIP (standard work-in-process). In this case, it would be 5.5 minutes times the maximum of 48 units in process (one for each workstation, and assuming zero queue in between any of them), which is equal to 264 minutes, or 4.4 hours. What we want to do is to reduce the number of stations, which effectively also reduces the SWIP on the line, and thus the amount of inventory tied up on the line.

Ken continued:

"There are many ways to do this; we'll go through a few of them here. One way is to add more operators on each station. For example, if we put two operators at each station (or zone, on a paced line), we would only need 24 stations (or zones), and therefore only 24 units of SWIP, still assuming there's one unit at each station. This would cut our production lead time in half, to 2.25 hours, and cut work-in-process inventory on the line in half. This would also reduce the square footage of space we need, cut the waste of transportation, and limit our capital investment. We could put three or more operators on a given station and cut the SWIP even more. Of course, there are limits. Sometimes there's not enough space for multiple operators, sometimes it's not safe, and sometimes the sequence of steps that can be done at a station will not support increasing numbers of operators. It's generally desirable to have more operators per station, and we should do so when the nature of the work allows it. However, there will be times when it's just not possible and we can only have one.

"Another approach is to pull process steps off the main assembly line and build them in feeder lines or feeder workcells that will, ideally, be located line-side. We might even move some of these to external suppliers, and ship them in, store them line-side in a 'just-in-time' manner. We call these high-level assemblies, or HLAs. If, for example, we could remove half of the process time needed on the main final assembly line and do this work offline, we could cut the number of stations or zones needed; in this case, probably in half again, as above. There are usually many other advantages to this. HLAs can typically be built in stations that are more ergonomic, with fixtures designed for a specific purpose. That means better task lighting and improved accessibility to tools, which ultimately means that the work can be done quicker and with better quality.

"Another thing we'll almost certainly do is to improve the processes on the main assembly line and reduce the times for each. If we could cut the total process time through kaizen by, say, 25%, we could potentially reduce the number of stations proportionately from 48 to 36. We could do the same by working with Engineering to revise the design, making elements more manufacturable and easier and faster to fabricate or assemble.

"We can and will do a combination of all these techniques, along with other tricks of the trade that I'll share later. This will be an 'evolutionary' process. As the details of the process and product design are worked out over time, we'll continue to revisit the question of the number of stations required and the best approach to take."

For the next stage, Ken asked the team to begin to design the process for final assembly by combining components, parts, or assemblies at different stations in sequence, and showing the work that would be performed at each station. He told the team to arrange the cutouts on the tabletops, but not to tape down anything, because they would move things around until they reached the point the team felt was the best.

How they could be sure they had it right? Ken told them not to worry about it. This was only the starting point, and their task was to do the best they could. They would continue to refine it and make it better as they went further into the 3P process. He encouraged them to consider reducing the number of stations and pulling as much of the assembly off the line as possible, but not to worry about the feeder processes just yet.

The team experimented with different possibilities again and again: having fun, feeling challenged and determined, occasionally arguing and debating, but, overall, lost in the exercise and oblivious to the second hand of the clock on the wall ticking further into the afternoon. By the end of the day, they had a design that they felt comfortable with (see Exhibit 7.1). Still, some worried that there was

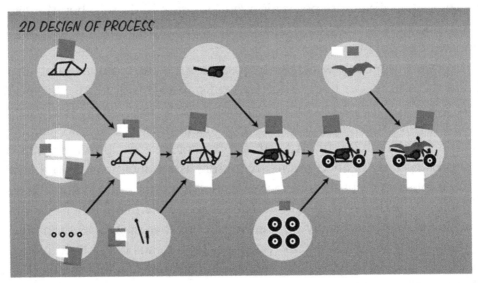

Exhibit 7.1   Initial 2D design of assembly process.

more they could do, Ken took a look at what they had, congratulated them, told them it was a great start, and that this was exactly where they wanted to be at this point. They adjourned for the day, each team member heading off in a different direction. Some went home, while some stayed and did other work. Each felt a strong sense of satisfaction. This was a lot better than their day jobs.

❋   ❋   ❋

The next morning, the team picked up where they'd left off. Several wanted to refine the main assembly line some more. But Ken encouraged them to hold their ideas. They could go at this forever and never run out of ideas, but they were in good enough shape, and later work would help refine the main assembly line. Instead, it was time to introduce them to a new concept: "Sequence 90."

SEQUENCE 90

Whatever our final process layout looks like, we want to develop a primarily sequenced system of flows, versus a mostly pull kanban series of flows, or worse, a series of push flows. I'll explain a lot more about this later, but let's get a high-level understanding of the goal of a Sequence 90 system and how to construct one. First, we start with the end-product when it's being loaded on a truck for shipping. At that point, we can say that it's worth almost 100 of COGS (Cost of Goods Sold)—the direct material, labor and variable overhead costs, less a few other costs like transportation, warranty, etc. The goal is to then walk back upstream through the build layers of the Bill of Material until we get to the point where our work-in-process and components downstream of that point are around 85 to 95 of COGS. In other words, we will have built up in our material and applied labor and overhead to about 5 to 15 of the total COGS in the Bill of Material. At that point, we would build a supermarket. This is where the "90" comes in in the Sequence 90, meaning about 90 of COGS plus or minus. So processes upstream of the supermarkets would generally be scheduled from pull signals out of the supermarket. The processes downstream would pick their parts from the supermarket and all be connected together. The main line would be the pacemaker and the schedule point. Production of all the feeder and sub-feeder assemblies would be controlled with sequenced signals from the main flow line, as would much of the externally supplied materials from vendors. This is ultimately how we want to design the process for the Trail Gripper.

"So, based on the Sequence 90 idea, we want to extend the final assembly portion a little more upstream than we currently are. Where we end it, we'll put a supermarket to pull from. That'll be the single-point scheduling location for the entire flow system. There are probably a number of choices. As we look at your current final assembly model, there are a number of sub-assemblies that will feed into it almost from the start. Any of these could be made part of the final assembly pacemaker line."

Ken gave the team an hour to experiment with different paths to best take the current final assembly steps further upstream. They asked for coaching along the way, concerned about whether they were thinking about it in the right way. Ken encouraged them to calmly and deliberatively work it through themselves, to talk about the pros and cons, that as a collective team they would come up with a good plan. He reminded them that this would not be the end design; it was meant as a high-level starting point only, and that they'd refine it over and over and make it better in future work. It was best to think about this as an initial rough design.

After about 45 minutes, the team decided that the best alternative at this time was to add frame fabrication after paint to the final assembly line. Ken took them through the same exercise as before when they estimated the total process time for final assembly using the data from the current Trail Gripper line. They came up with a total process time of 3,960 seconds for frame fabrication. This translated to 12 additional operators and 12 maximum additional stations/zones. Ken asked them to spend an hour and a half on these 12 stations, applying the same guidelines for reducing production lead time as they'd used earlier on the rest of the final assembly line. When they'd gone as far as they felt they could, Ken commented that they had a good crude initial proposed layout for the main assembly line and sent them off on a break.

After the break, they spent the rest of the day on their 2D process layout, going through the exact same task, trying to figure out where to connect the various feeder lines into the line, and how their flow would be constructed. One team member commented,

"This is like a jigsaw puzzle."

Ken responded, "Yes, but you get to design the pieces of the puzzle as you put it together."

Toward the end of the day, they had built what looked like a real flow system, with scattered cutouts of part and assembly drawings mixed with hand sketches of the same. They were proud of themselves and what they'd managed to accomplish in just two days. They actually had a fairly detailed concept of how the product would flow, including specific parts and sub-assemblies. More importantly, people from different departments were now involved in developing the concept. In the past, their ideas would not be heard until much later in process—if at all.

❈   ❈   ❈

The next morning, Ken took attendance. With the entire team on time, present, and accounted for, he launched into the next exercise. He broke up the team into groups and had them take on different sections of the line. He challenged them to come up with process times for each operator and to write this information on a Post-It® note—one note for each operator, with the total process time listed—and to hand them in as they went through the exercise. He asked them to try to break down each total process time into five to ten steps of around 30–60 seconds each, and to jot these sub-steps and times on the Post-It® note as well. This wasn't critical for each station and operator yet, but the team should do it where it made sense.

Scott Green, the material controller, was confused. He asked how they could possibly do this, as all they had at this point was a very general concept of what work would be performed at each station. Ken told him it was a great question, and reiterated that this was a first pass and meant as a rough estimate only. Again, they would refine these times over and over throughout the project. The intent of the first pass was to examine two main things: the operator times that were estimated to be under the takt time and those that were estimated to be over the takt time. He told them that they could start with times from Trail Gripper and use similar times from their own operations. Barring that, they could bounce estimates off each other based on what they saw the process as likely to entail. He reminded them not to spend a lot of time on it or to worry too much. Whatever they came up with, they would be working to make more accurate and reduce anyway over time. The team got started.

> Johnny Cox, the shop assembler, said, "Hey, can we play 'air guitar' to work out the rough estimates?"
> Ken smiled. "That's a great idea. Anything you can do to make the 'fuzzy' more tangible is always a good idea."

He pulled John and Gina, the two Lean leaders, aside, and told them they would be assigned to a different task. The same held for Dave and Pete, who would also work on something different but related.

He asked Pete and Dave to grab a bunch of butcher paper sheets and start to write across the top of the sheet in pencil a section for Current Hours with a column for Process Station, Operator Number, Step Times, and Total Time (Exhibit 7.2). They would be developing a Process Planning Time Sheet. They needed a section beside that for New Base Line with columns for Step Times and Total Times. He then directed them to start listing all stations on the final assembly line in the left columns, then to add the stations in the feeders and sub-feeders one by one. When there were multiple operators at a station, Ken instructed them to list each operator, leaving enough room under each one to list up to ten steps or rows. When the teams submitted their Post-It® notes, they were to record the information under the Total Time column beside each operator/station; and if they had a breakdown of sub-steps, to list these beside each operator in descending rows listing their times. Dave and Pete got started.

John and Gina would be building a Yamazumi chart (see Exhibit 7.3). They got some butcher paper and taped four sheets together end to end, knowing that

Process Planning Time Sheet							
Current Hours				**Target Hours**		**Sensei Target Hours**	
Sequence Element	**Work Area**	**Sub Hours**	**Total Hours**	**Sub Hours**	**Total Hours**	**Sub Hours**	**Total Hours**
Assembly Line			77				
	Packing	1					
	Test	1					
	Main Assembly	57					
	Frame Assembly	18					
Material Handlers			12				
Sub Assembly	Engines	55	131	➡		➡	
	Transmission	51					
	Axle/Wheel	10					
	Dash	15					
Total Main Assembly			220				
Frame Fab			35				
Component Fab			63				
Paint			4				
Thermoform			8				
				Delta	-66.0	Delta	-39.6
Total Hours			330		264.0		224.4

Exhibit 7.2   Process planning time sheet. (*Source:* Shingijutsu.)

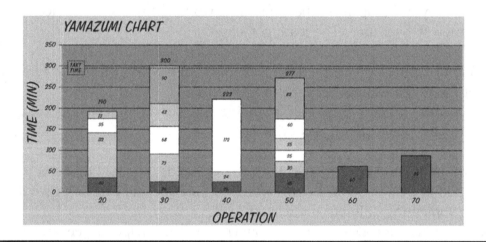

Exhibit 7.3   Yamazumi chart.

if this wasn't enough, they could easily add more later. Ken instructed them to draw a line across the bottom with enough room underneath for labels, and then to draw a vertical line on the far left with enough room beside it on the left for labels. The vertical line would represent time. They then broke down the vertical line into eight equal segments, labeling them from 1 to 8 minutes. Next, they drew another horizontal line in red at the 5.5-minute line, representing the takt time. Along the bottom, they listed the stations on the main line, the feeders, and finally the operators. If there were multiple operators per station, they were to line the station label under those operators, each operator about the width of a Post-It® note. They would build a stacked bar of Post-It® notes, each representing a specific step and that was cut to a height that was proportional to the process time for that step. They would use the scale they'd already created on the vertical axis to determine the approximate height. The goal was to show a rising progression of Post-It® notes, so that its total height equated to the total process time shown on the Post-It® note on the Process Planning Time Sheet that Dave and Pete were constructing. John and Gina were now prepared to create the Yamazumi chart.

The teams chipped away at the exercise all morning. By lunch, they had completed their first pass. They gathered around the Yamazumi chart to study their results.

Ken asked them what they saw. The team looked at the stacked bar of Post-It® notes for each operator and how far each stacked bar rose vertically. They could easily see how each compared to the takt time of 5.5 minutes. They pointed out that, of the 60 operators on the main line and additional 40 operators in the feeders and sub-feeders, 15 of them had times that were lower than 50% of the takt time, 10 of them had times between 60 and 70% of the takt time, and 16 of them were over the takt time.

> "Great work, team. This is good enough for now. Although these numbers aren't necessarily accurate, we can already see some potential problems with line balancing. We'll likely have to work to get all operator times that are above takt time, below it. We also see some operators who don't have much process time as compared to takt time. We may want to combine their work with other operators to bring them closer to takt time. That way, we may be able to reduce the number of operators in the system and the number of stations, which will help us shorten and simplify the flow. As we move forward, we'll work on all these challenges and opportunities, and we'll maintain and update the Yamazumi chart and the Process Time Capacity Sheet. That's why we've done them in pencil and with peel-off Post-It® notes. We'll be updating them often, and that'll make it fast and easy. Let's grab some lunch."

Before they left, George Hall, the Advanced Manufacturing Engineering Leader commented,

> "Wow! We wouldn't have discussed line balancing for months if we designed the line the way we used to. And this is what, Day 3?"

Everyone nodded in agreement, and they went off for a well-deserved lunch.

Pete was pleased with how engaged everyone was. Of course, maintaining the momentum over the coming months would be a challenge that would require his steady leadership. Dave was also thinking. It had struck him how little he really knew about 3P. It wasn't just a collection of concepts, such as "air guitar," but a robust process. He shouldn't have been surprised by this. Every aspect of his personal Lean learning journey had involved a process, and he couldn't wait to see how the entire process would play out.

❋ ❋ ❋

In the afternoon, the team reconvened for a new exercise, focusing on one of the most powerful parts of 3P.

"Process-at-a-Glance" said Ken, "is the next phase of our project. Now we'll really start mapping out the details of how we'll manufacture the Trail Gripper."

Ken rolled out a blank template on one of the tables, and with Mary's help, taped it to the wall.

"Let's gather around, everyone, and I'll take you through this tool." (See Exhibit 7.4.) "We're going to run through how Process-at-a-Glance works. We'll use it two different ways. First, we'll split into four teams and break out sections of the flow system from our diagram, assigning them to each team. Each column on the sheet is a progressive listing of each sequential station in the process. Second, where a given station has a number of significant work-in-process transformations, you may opt to use consecutive columns for each major transformation in a given station. In other words, we'll show the consecutive build in these stations.

"So, if we add Station 1 in the left column, you'll go through and sketch out the best idea that your team can come up with for each category that the row represents. We don't need to worry about being precise, get into tremendous detail, or be a great artist; just sketch out the concept in a simple way that conveys the idea. The first row is for a simple sketch of the part or work-in-process assembly that's at that station. Be sure to include all parts that you feel need

Process-At-A-Glance											
Part Number _____				1. Takt Time 2. One Piece Flow 3. Pull system				_____ _____			
											Sequence
											(1) Material process sketch
											(2) Process method
											(3) Poke Yoke No-Go Gage
											(4) Tools
											(5) Jig or Fixture Hanedashi
											(6) Machine

Exhibit 7.4   Process-at-a-Glance: Blank. (*Source:* Shingijutsu.)

special processes, or special quality steps, or have a significant impact on the customer. In other words, focus on the important stuff. Keep in mind that you'll probably want to erase some of these as better ideas play out, so let's do all of our work in pencil. If you think that a number of transformations will be performed at the same station, sketch out the part or assembly being worked on for the first phase of the transformation at that station.

"The second row is for a sketch of the process method that will occur at that station. How will the materials be transformed, and by what process methods? Just draw a sketch of the primary process method that the team believes is the right and best way to do it. Keep in mind that we should only show the primary method. You don't need to try to show every activity. For example, if the primary method focuses on how to get the motor transferred in and mounted into the frame, we should sketch how we'll do that. If there are four bolts added, we wouldn't show those secondary operations. We'll plan for those when we get to the mock-up phases of the project.

"We'll use the third row to think about how we'll mistake-proof the process method at that station, particularly if the prime operation is critical to quality. In the fourth row, we'll sketch out any of the major tools we'll need at that station to perform the process method. In the fifth row, we should sketch any jigs or fixtures that will be needed; and in the last row, if the transformation is performed using a machine or machines, we should sketch that out.

"Here's a key point. The main intent of this exercise is to just start to understand the elements of the process itself. The best way may not always be apparent to us right away. There may be methods used on the current Trail Gripper line that we think there's a problem with. If we can't come up with a way, we'll add a 'bomb' signal in that box on the chart. If an idea comes later, we can remove the bomb and add the sketch. If not, we'll use a 7-Ways process, which we'll discuss a little later.

"Any questions?" asked Ken.

"Seems pretty straightforward; why don't we get going and see how we do?" Dave said.

Ken split up the teams and assigned each of them a section of the main line. Dave, George, Mary, and Bill Cook were on one team and were assigned the second half of the main line. They chipped away station by station, following Ken's advice. At first it was a little difficult and confusing, but as they detailed more and more processes, it became easier, and they went faster. By the end of the day, they filled out their first Process-at-a-Glance sheet, with 11 stations in sequence. They did have a few bombs, as shown in Exhibit 7.5. Ken had them post their sheet on the wall and prepare to start a new sheet, picking up where they'd left off on the prior one.

All of the other teams chipped away as well. The process took almost two weeks, but they made it through all the stations. The walls were filled with

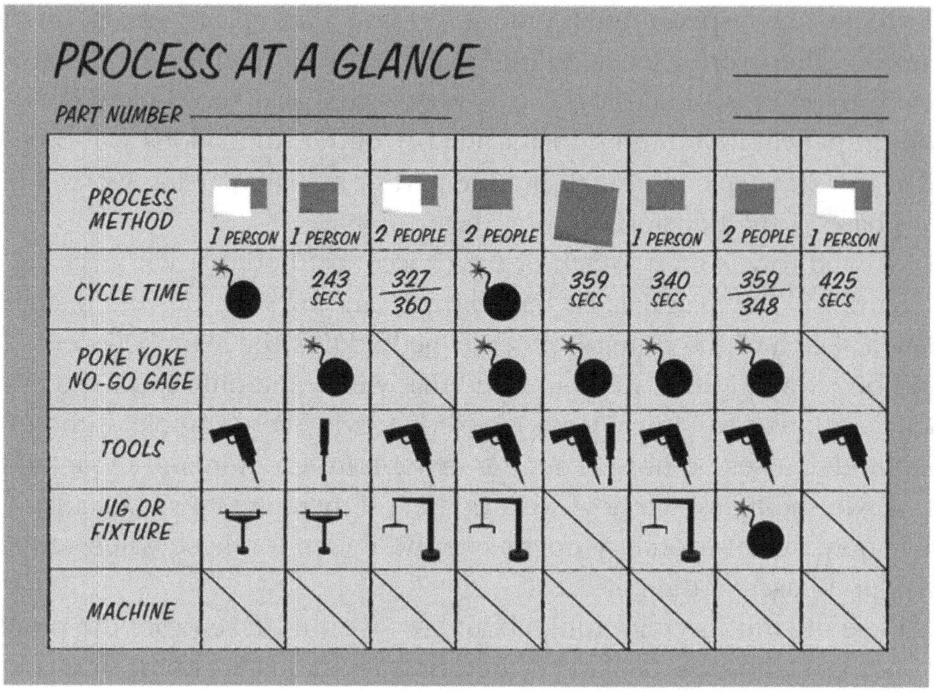

Exhibit 7.5   Process-at-a-Glance: Filled in. (*Source:* Derived from Shingijutsu.)

Process-at-a-Glance sheets in order of flow. Ken reminded them for the tenth time that the Process-at-a-Glance is a living and breathing document; in other words, if a better idea came along, they'd have to update the sheet. He also told them that as they got the results from the upcoming 7-Ways work, they needed to have the discipline to remove the bombs and update the sheet appropriately with the results of the 7-Ways exercises that they'd soon be conducting.

❋   ❋   ❋

The team assembled the next morning, ready for the 7-Ways exercises that Dave had described to them. Sure enough, Ken told them they were ready to solve some of the challenges that had plagued them during the Process-at-a-Glance. They took morning attendance and had their ten-minute morning meeting, where they reviewed their progress against the schedule that they'd developed weeks earlier. They also reviewed issues from the prior day that they hadn't resolved and how they would tackle them, and discussed the key objectives for the day ahead.

Before they got started, Ken planned to take them through a 7-Ways training program. Just as he was about to begin, George Hall asked if they could begin to lay out some of the process steps. He said he was excited by what they had to date, had been thinking all last night about a number of ideas, and wanted to get going. First, Ken acknowledged George's enthusiasm. He then spoke to the entire team and told them that there will be times that other team members will want to do similar, to skip a few steps in the 3P process. However, he was going to walk them through a specific process that they would repeat over and over to develop the layout, the workstation and equipment designs, the standard work, and even elements of the product design itself. He said that everyone must trust the process.

TRUST THE PROCESS

In 3P, to design the elements of our process and elements of the product design itself, we'll go through 7-Ways exercises, develop Process-at-a-Glance sheets, mock up all elements, simulate them, observe the simulations for waste, go back and kaizen the mock-ups, and simulate and observe over and over until satisfied that we have the best concept we could currently come up with. We need to avoid the very common human desire and behavior to jump to solution. We need to trust the process—trust that if we take the time and do the work, it will always get us to the best answer.

Ken explained that as their coach/facilitator, he would stay on them to stop jumping to solution and do the work, as annoying to them as that might occasionally be.

"We're going to start getting into the meat of our process and learn a very powerful exercise that will not only help us develop a world-class process for Trail Gripper, with simple, easy, and safe work for our operators, but also one that we can use any time to continuously improve any aspect of any process, as well as the products they produce.

"I want you to consider the flow of the process. As you look at a landscape, all you may see from a distance are the mountains. As you get closer to the mountains, you see the forest running up their sides. As you look forward further, you start to focus on an individual tree, and as you get close enough, you can actually touch it. As we start with a high-level process design, it's like looking at the process from 60,000 feet. The Process-at-a-Glance and 7-Ways take us down to the 5,000-foot level, where we focus on specific elements like tools and fixtures. As we move from benchtop-scaled mock-ups to full-scale ones, we get to see the smaller elements more clearly. Ultimately, when we get right down to developing production tools and fixtures and such, we can actually touch and feel them. This is the essence of the journey we're on, and the 7-Ways process begins to get us closer to the real elements of the system that we'll develop.

"So here's what we're going to do. We're going to break into three teams of five to six people each. Each of the teams will go into separate breakout rooms, and over the next week or so, we're going to take each of the bombs from any of our Process-at-a-Glance sheet elements and we're going to go through a 7-Ways process on it.

"Here's how it'll work. You'll get together in your breakout room with an easel and a bunch of Post-It® notes and some pencils. Each one of you will sketch a possible method for doing the task or tasks at that station. Remember when we talked about the need to remember what it was like when you were younger and to learn again to think like a 12-year-old? We need to do this here, all the time. We want everyone to be totally open-minded. No idea or way is crazy; sketch them all and put them on the easel as you get some done. Don't worry about fancy or good sketches; that's not important. Just draw your idea the best you can. When you post your ideas on the easel, take a minute to look at the sketches others are doing, as they may trigger totally new ideas in you, even wild and crazy ones.

"There are a number of steps in the thinking process that I've put up on the wall to remind us (Exhibit 7.6). First, you need to determine the function or functions that need to occur at the station. Determine the theme and scope of the work to be done at that station. Is it on the main line or in a workcell?

"Second, think about the essence of the function itself. What are we doing here? Are we doing a turn, a spin, a fasten, a press, a shear, a drill? What's the essence of the function? The next part may be difficult for some of you. It may seem weird, or uncomfortable, or maybe even dumb. However, this is perhaps the most important and powerful trick in creating great processes. Once you determine the essence of the function, we want you to think about an example in nature that most typifies it. It would be good as you start each 7-Ways to discuss the essence of the function and the example in nature with the team before you go on. So, for example, what in nature might spin?"

Brenda shouted out, "A tornado!"

Gary said, "A whirlpool!"

"Very good; both are excellent examples. What in nature has a hinge?"

Mary shouted out, "A shoulder ball and socket!"

"Excellent! I think you get it. The reason we want to think about this is that nature often has developed the purest of ways to do anything. If we think about and study an example, it can help us think of ways we might do the function in a similar manner.

"Before you start sketching an idea, sketch out examples from nature, study them, think about the phenomenon at work and *how* it works, how it connects with other things. Then sketch the details of how it actually works. In the case of Mary's ball and socket, think about how the shoulder actually attaches and moves. What are the different elements that enable all of this to happen? Remember: Think like a 12-year-old and then start to sketch some man-made ideas to emulate the same function that nature does. Each team member should try and develop seven different ideas themselves. That's why we call it '7-Ways.' Be creative! Stretch your imagination!

Exhibit 7.6   Steps of the 7-Ways process. (*Source:* Derived from Shingijutsu.)

"Once you've completed as many sketches as possible, the team needs to try and group them together under common themes. Arrange them all on the easel. If you need more space, put them on multiple pieces of butcher paper and tape to the wall (see Exhibit 7.7). We'll save all of these for a while in case we feel a need to go back and explore a different idea.
"Any questions so far?"

None were raised.

"Once we've done this, the team needs to vote. You want to try and get down to the best three by consensus. Decide as a team how you want to vote. I've found the way that tends to work the best is for each member to rank their top three choices in order of preference: one, two, and three. If there are a lot of sketches, you might even do your top five. Sum up which idea has the most ones, twos, etc., putting more weight on the higher preference. Usually, the top three will jump off the page this way.

"When you have your top three, I want you to mock up each idea in benchtop size."
Dave spoke up. "What does that mean? How are we supposed to do that?"
"I want you to mock them up into a small-scale 3D model. You can use anything for this. Get creative. Cardboard, paper, Styrofoam® pieces and cups, tape, balsa wood, straws, wire, steel, copper, whatever you can think of to build a model. Make them as true to the idea as possible, but a crude mock-up done quickly is much better than getting too detailed and taking a long time. You'll be building many mock-ups with the materials we gathered weeks ago.

"Our experience at other companies has shown that a real 3D version allows people to perceive nuances of the design that a 2D sketch or drawing, or even a 3D modeling program, can't do. Modern 3D modeling programs are invaluable for helping designers get better quality, manufacturability, and functionality. We see mock-ups as a powerful complement to 3D development software—not a replacement. And the average person can see and understand the workings of a physical mock-up, as well as the problems and wastes associated with it. Staring at an engineering model on a computer screen just isn't as easy for most people to do."
"What do we do with the mock-ups?"
"George, you'll use each of the mock-ups to simulate how the work will be performed in the station. You'll want to mock up the parts that will be used to make the simulation more real and intuitive. You'll learn a lot about what works and doesn't work through these simulations, and develop better and better ideas along the way."
Ken continued, "As you're doing this, you might identify some changes to the product design that you want to recommend. Each team should rely

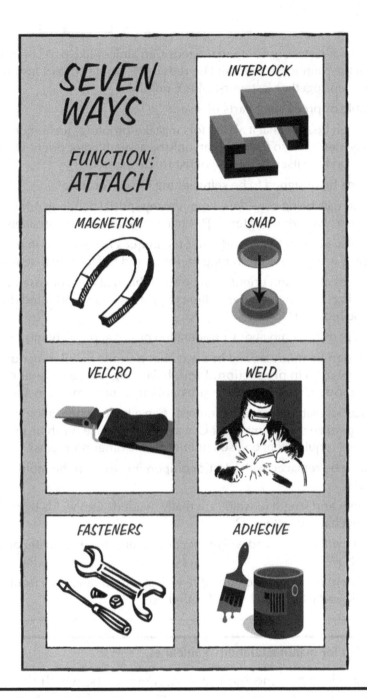

Exhibit 7.7   Results from 7-Ways process.

on Brenda, our product design engineer, to review any suggested changes. This is a good time to review several guidelines for 'design for manufacturability' that you should consider as you go about your design effort. Keep in mind that this is a partial list of questions to prompt your thinking. You may come up with other ideas on your own."

Ken posted the preprinted list on the wall and proceeded to review the list with the team (see Exhibit 7.8).

- Can the type of fasteners be standardized? Can almost-like parts be made the same? Can the number of fasteners be reduced? Avoid two-part fasteners. Use captive or snap-together fasteners, where possible.
- Can available or preexisting parts be used?
- Can the design be such that it allows for "mistake-proofing" techniques in assembly? Can parts be designed so that they can only be assembled or oriented in one way? Will the design lend itself to the use of fixtures in assembly?
- Can the parts be designed to be self-aligning or self-locating?
- Will the materials being considered require any special handling or machining processes (e.g., hazardous materials)? What will be the impact on safety? Cost?
- Can slots, pear-shaped holes, and other techniques be included in the design to permit use of less-precise machining methods and to facilitate assembly?
- Can part accessibility and orientation be such that it allows for ease of manufacturing? Can the number of times a part must be reoriented during assembly be reduced?
- Can several parts be combined? Consideration of size and weight must be given.
- Can the design be modularized in a way that permits cellular manufacturing to be effectively used in production? Work balancing is a consideration here. Ideally, no module should require substantially more time than another.
- What "grade" of material is really required? Can a lesser grade, alloy, etc. be used and not jeopardize design integrity? Can a coarser surface finish be used? Can a lesser flatness requirement be used but not create other problems?
- Can a casting be replaced with plastic components that can be molded, extruded, or formed?
- What number and length of welds are really needed? Can welds be replaced by bends or mechanical fasteners?
- What impact will the size and weight of components have on material handling in operations?
- Are component dimensions compatible with standard raw stock from which they will be made and which can be easily purchased?

Exhibit 7.8   Design for manufacturability guidelines.

"This list is applicable to mechanical design elements, which the vast majority of our work on the line represents. There are separate lists for electrical design elements, software design, even packaging. We'll keep our focus on these for our purposes.

"Back to simulating. The rest of the team should act as observers and make notes about what they like and don't like. I need you to record the strengths of the design and look for wastes and problems. Quality, mistake-proofing, and safety are major criteria to look for, as well as ease of use, speed of use, simplicity, potential reliability, and ease of maintaining. We want to look at any required changeover times and how well they support one-piece flow. We also want to consider cost and ease of production, but keep in mind that our costs will be a lot lower than they might be traditionally. Once we go through the three simulations, you should get together again and rank them. I can help you with this if necessary. Take everyone's input, rank-order them, and pick the best design. Once you do, we'll lock it in. Does anyone have any questions about this process and the expectations?"

"Yes, I have a few," said Dave. "What do we do with the final choice? Do we build a prototype? And what do we do with all of the mock-ups?"

"We'll go on to put Process-at-a-Glance sheets together, where we look at all elements of the design, and from there we'll move on to build a 1/5 scale mock-up of the benchtop model for further simulation. We should also build a simple clean storage area to save the benchtop mock-ups for future reference. If we get far enough along, we may throw out the two unused designs in each to save space and keep the selected one or the ones that we feel are most important. At times we may need to go back and reference them. If there are no other questions, let's get going."

The teams worked diligently every day for three weeks, doing numerous 7-Ways exercises. They produced mock-ups and went through simulations every day (see Exhibit 7.9).

Brenda, working with several teams, performed 7-Ways exercises on aspects of the product design as the need arose. One resulted in a unique method of attaching a component to the frame using a simple interlocking mechanism rather than assembling with multiple fasteners or more complex processes such as welding. Productivity was very high, and Pete participated as much as he could. He made a point to come in every day for at least an hour. Bit by bit, the team got through the models and had a real benchtop layout of the preferred designs for much of the flow system they were creating. Where it was obvi-

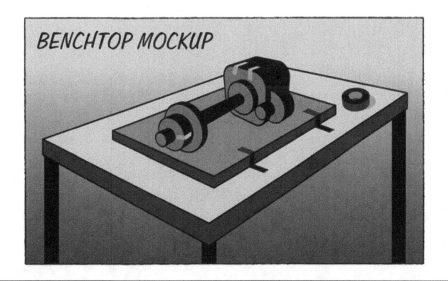

Exhibit 7.9 Benchtop mock-up.

ous a process time had changed from the original estimate, they updated the Yamazumi and Process Time Capacity charts.

❋   ❋   ❋

Pete had received news that Steve Sawyer was going to come to town in a few weeks. Pete had given him an update, and Steve wanted to get a first-hand look. Pete was at home sitting in his family room when his wife Debra asked,

"What are you thinking about?"

"Oh, just thinking about one of our executives coming for a visit."

"Is that worrying you?" she asked.

"No, not at all. Just the opposite, in fact. I'm actually looking forward to it. We're doing really good work now. The 3P process that Dave introduced us to and that Ken is leading us on is amazing. We're actually developing all of the processes for this new product we're bringing in, down to the smallest details, well ahead of the transfer, and the level of engagement of the folks we have involved is higher than I've ever seen. I think that we're learning a game-changing new way to drive major change, and I think that Steve is going to think the same thing when he sees it."

Debra said, "Well, I've never seen you excited about having your boss come to visit before. Usually, you'd be nervous, putting last-minute touches on some presentation you would be making. You seem…"

"You're right. In the past, a visit from the boss would have meant trouble. This time, it's different." He paused, smiled, and said, "I'm not even going to put together a presentation. I'm just going to take him out to the area where the team has been working and he can see for himself. No fuss, no muss."

❋   ❋   ❋

Chapter 8

Simulate–Observe–Kaizen–Repeat

Pete swung the rental car across two lanes of traffic into the entrance to St. Lucia's Hospital and curved around the circle and into the parking garage, finding a convenient spot right on the first floor. CEO Frank Kent had come through, as promised. Frank had contacted CEO Joan Jarrett to arrange for them to benchmark St. Lucia's 3P program. What's more, Joan herself was waiting in the lobby to greet the team. Pete had brought five other team members with him: Dave, Ken, Mary, George, and Brenda.

Joan was the consummate ambassador and enthusiastically greeted each team member. Pete thanked her and the hospital for going out of their way to share their learning with his team. Joan waved him off and said,

> "It's no hardship at all. We're still relatively new to this 3P process as well. I understand that Trail Gripper is also approaching it in a big way, and perhaps there would be opportunities in the future for you to reciprocate and let my team come visit you. Let's get started."

Joan led them through a maze of sterile, well-lit hallways. It was obvious that Joan was both proud of and intimately familiar with her institution, as she pointed out and described one department after another.

Finally, they entered a spacious conference room, where three St. Lucia's employees were waiting for them. Joan introduced them to Megan Taylor, one of their doctors; Virginia Hall, one of their nursing supervisors; and Amanda Stewart, one of their nurses. The two teams shook hands all around, and Joan explained that Megan, Virginia, and Amanda were key members of one of their more successful 3P programs and would be walking them through their process and answering any questions they had.

They walked as a group to take the elevator to the basement of the building. Turning right, they entered a large open space of well-laid-out mock-ups and models, walls lined with charts that looked remarkably similar to those in the

Trail Gripper team room that they had just left yesterday. Pete could not help but be amazed as he gazed around the room. Despite how much different their processes were from this hospital's, here they were with a 3P process that could have been a carbon copy of theirs. St. Lucia's was further along in their process, but their 2D flowchart was right there, as large as life, looking much like their recently completed layout depicting the proposed Trail Gripper line.

Megan started out by explaining that their entire objective was to improve patient care. "When we looked at the current state of one of our general units, we realized we need to change the way we were organized to fundamentally improve care to our patients. We wanted to create better care process flow to eliminate much of the waste that we mapped out. There is a lot of motion waste in our existing units. Nurses in particular do a lot of running around. We identified a lot of waiting waste, waiting for meds and needed supplies for example. We also wanted to improve the visibility that the staff has of the patients. Right now there isn't much 'line of sight' given the poor layout, which is basically just a very long hallway (see Exhibit 8.1). This creates more running around just to make sure that everyone is OK. You can see what we are looking to do to address that. We've decided to go with a concept of a circular layout divided into four quarters. Rather than build actual walls, we mocked up where they would be so you'll have to use a little imagination. Each quarter is an exact duplicate of the other, so we built full-scale mockups of all of the equipment and supplies in the one. You can see that we have five rooms arced around within each quarter, with two beds in each room for a total patient load of ten patients per quarter.

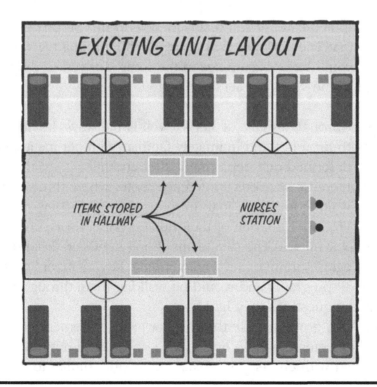

Exhibit 8.1 Existing unit layout.

Each room is equipped for each patient's needs. All things required for patient care are either included in the design of the room or delivered directly to the room using a 'pull/kanban' system. With four quarters, the total unit would house 40 patients."

Virginia went on to explain the design of the area outside the rooms, which was mocked up using wood and large foam exhibits. "Given that everything needed is at 'point-of-use' or delivered directly to the room, we were able to eliminate the need for a nurse's station. The mobile carts that you see mocked up will be what the nurses use. The carts will be equipped with wireless devices as we move to a paperless system. Over here is our support center located in the middle of the four quarters (see Exhibit 8.2), and will include everything it needs. Unlike our current model, this support center is cross functionally staffed with several support services. The doctor on duty will be located in the middle of the room along with the other support staff such as a pharmacist, and a phlebotomist who are serving all forty patients in the unit. It represents a collaborative care model that we are striving to implement. The support staff will vary

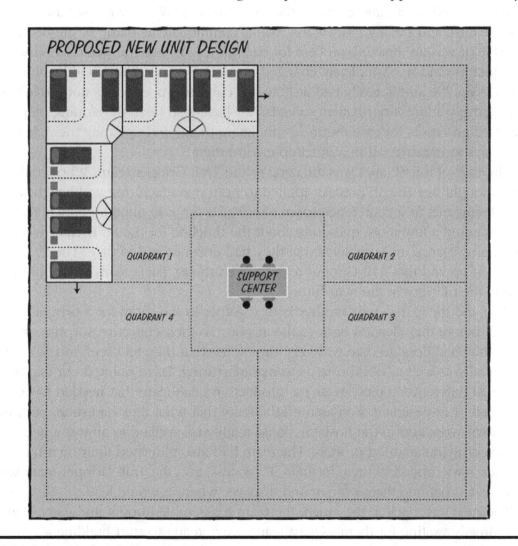

Exhibit 8.2 Proposed new unit design.

based on the type of unit it is. The staff required for a Labor & Delivery unit would be different from a Critical Care unit. This is going to be a major change for us, as these resources are now located in their own departments, often far away from the patients. Today the various support staff periodically make their rounds and usually by themselves to see the patients. Now, they'll be right here in the center of the unit. We've been simulating the workflow for a month now, taking waste out and creating standard work for each role. Working as a team, the doctor and the support staff together will walk the unit at least three times a shift and support each other in real-time to serve patient needs."

Amanda walked them through one of the patient rooms, showing them how the room was fully structured for both patient care and to provide a pleasant environment for the patient. The room included all of the mockups, just as the central support center had been, including cabinets for supplies and other items.

"We've all taken turns acting as a patient, lying on the beds, and have charted numerous needs and desires. We even brought in several types of beds and each of us took turns spending an hour in each one, and rating what we liked and didn't like, both as a patient and as the care provider. We're working hard to find cost-effective and quality models that will accommodate as many of the needs as possible. It was an amazing process for each of us. We see the world as nurses and doctors, and we completely changed our mindset to see the world as our patients do. We all felt really bad at how our current process does not enable us to effectively address our patient's needs. We have also been developing more standard processes for care routines, simulating them, and updating them as we find ways to improve, all in a mockup environment"

The tour of the 3P area was amazing to the Trail Gripper team. Who would have thought that the 3P process applied to both manufacturing and healthcare? Both teams met in a conference room afterwards for a 30 minute debrief. The visitors asked a few more questions about the timeline for the St. Lucia project, and the resistance and problems that they had encountered to date. They wrapped up and the Trail Gripper team offered the St. Lucia team an opportunity to visit their facility in the near future.

Pete and the team spent another hour together in the room for a private debrief before they headed back to the airport. To their collective surprise, the work they had done to date was very similar to what they had seen today, applying equally as well to healthcare as to manufacturing. Dave pointed out that the only real difference is there wasn't a 'product' in healthcare that needed to be designed or re-designed. Ken assured the team that what they had seen, both in their own work and in the hospital, could readily be applied to almost any office, service or manufacturing process. The team had also glimpsed their own future. St. Lucia's was ahead of them in their 3P process, and the Trail Gripper team was anxious to continue their journey and discover where it would lead.

The next day, back at the ranch, the team reassembled bright and early. Ken was already waiting for them. "Okay, team; we're ready to start building a 1/5 scale model of our new process design."

Moonshiner Bill Cook looked stunned. "You mean the entire factory process for Trail Gripper?"

"That's right, Bill. That's exactly what I mean,"

There were a few soft whistles around the room Ken split the team up into four groups. Two groups would each work on half of the main line, and one would take the subassembly lines. The other team would focus on the conveyance method for the main line, and then spin off and help with the subassembly lines. Their goal was to take the conveyance method developed through the 7-ways and Process-at-a-Glance phase, and using their early bench-top mock-ups, create a 1/5 scale model. They would then run through a series of simulations and kaizens to further refine the design and ensure that most of the wasteful elements were eliminated. Brenda Lewis, their product design engineer, would be available to all of the teams on an as-needed basis, with each team asking her to review any ideas that might affect the product design. They were given a target of three days to refine their designs, and asked to run through at least three full simulations. At that point, they would hand their designs over to the two teams working on the main line, who would integrate the model for the conveyance method with their models.

Dave was teamed with George Hall, Mary Long, and Bill Cook and assigned to the back half of the main line. They started with the station that was marrying the engine to the frame. Like the conveyance team, they started with the bench top mock-up and built it into a scale model. They opted for a wood structure to simulate the design of the A-framed shaped device that they would use to move the engine over from the feeder line across and into the frame. They then used cardboard and foam to depict the other elements of this station. For example, they mocked up the air tools that they would use, as well as the air lines and balancers that would be needed. They did this with twine and cardboard cutouts of balancers and tools. It took them about two hours to design this first station, discarding a few designs along the way. Once they got through a few stations, the pace of the work began to accelerate.

Ken came by to look at their first scale mock-up and congratulated them on the work. He encouraged them to build a scale model for the next three stations, and then come to see him, as he wanted them to start a round of simulations. Using the same method, but moving a little faster now, the team completed two more stations within the next three hours.

By this point in the project, Pete had gone back to his day job, albeit very reluctantly. He had a plant to run, but he had enjoyed the work he had done on the 3P program immensely. His knowledge of lean and of process excellence in general had skyrocketed, and he really wished he could have stayed with the team throughout the entire process. He did however modify his leader standard work that Dave had encouraged him into starting all those many months ago. Pete had eventually come to love the plan that he had developed for the key activities that he as Plant Manager has to perform, each day, week or month. The plan in the form of a checklist served as an effective reminder to Pete, and compelled him to more consistently perform these activities. He vowed that he would never again work without it. After all, if he expected everyone else to follow standard work practices, he as their leader has to do the same. Therefore, he added a 15 minute check-in every day with the 3P team, with a full hour every Tuesday and Thursday afternoon. This will insure that he is giving the project sufficient attention.

On Thursday, Pete poked his nose into the 3P area and observed a beehive of activity. Four independent teams were all at work, constructing what looked like a miniature movie set, with Ken shuttling between them coaching, guiding and encouraging. Brenda Lewis was working with one of the teams, along with another person he didn't recognize. Pete walked toward them.

"Hi, Brenda. Who's your friend?"

Brenda responded, "This is Ben Smith, from the Product Design department. I couldn't keep up with all of the requests from the teams for product design enhancements so I asked our manager if Ben could help out. He said that I could bring Ben along, so here he is. I'm just getting him up to speed."

Pete remembered the conversation he had with Brenda's manager months earlier, pleading with his peer to include Brenda on the team, if only part time. Pete responded, "The more the merrier. Welcome to the Trail Gripper team, Ben." Ben and Brenda returned to the discussion that Pete had briefly interrupted.

Pete walked over to Dave's team, and Ken came over and told Pete he was just in time to observe their first scale model simulation. They moved over to the mockup of the first three stations of the back half of the line. Ken split them up by task. They started with the engine mount station (see Exhibit 8.3).

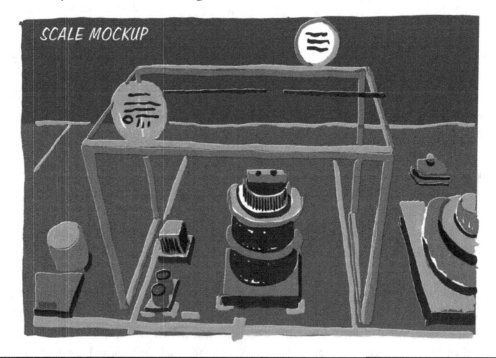

Exhibit 8.3   Scale mock-up.

Time Observation Form

Task / Process Being Observed:		Observer(s):								Date & Time of Observation:			Page Of	

Step No.	Task Component	Observation No.										Assigned Component Time	Remarks
		1	2	3	4	5	6	7	8	9	10		

Exhibit 8.4   Time observation form.

George would act as the operator and simulate the first station process, in which they would move the engine over from the feeder line and mount it. Dave would record the steps and the time to complete each on the Time Observation form (Exhibit 8.4). After they had conducted this simulation, they would do the other two stations as well, and insure that they could successfully meet takt time at all three stations.

The station included a mockup of the engine and the frame with all of the other assemblies on it up to that point of the process. George started by sliding the engine across the transfer arm, which was supported by the A-frame shaped device over the station. He lowered the engine onto the mounts through the cup and cone they had added to the fixture. Dave recorded all of the steps on the Time Observation form, while Bill used the stopwatch to time the movements, and called out the times to Dave. Mary and Pete looked on with a pad of paper on a clipboard to record any waste they saw. Ken stood nearby advising them on a few things to look for. On the first task, Mary noted that George was on the wrong side of the assembly and had to reach and strain to maneuver the engine over. Pete noted that while the cup and cone allowed for fast and easy alignment with the engine mounts, George had to lower the engine down two feet with the A-frame suspending the transfer arm. This could have been lower to begin with.

George simulated bolting the engine onto the frame with four imaginary bolts using the cardboard air driver that was suspended from a balancer. He simulated the connection of a harness, and then a conduit that he walked three feet over to a cardboard point-of-use material storage rack to retrieve. Mary noted that the rack could have been a bit closer to the station, and the shelf that the conduits were stored on could have been a little higher, as an ergonomic improvement. Bill continued to start and stop the stopwatch, calling each step's time out to Dave, who recorded them. Next, George stretched two belts, also taken from a rack on the side of the line, into place. He placed a label on the engine and, using a cardboard bar code reader hanging on the side of the point-of-use shelf, scanned the bar code on the frame and the engine to simulate a transaction necessary for their quality system that will allow future traceability of the specific engine to a specific frame in case a quality issue arose. Pete noted that George had to take two steps, stretch and reach to get the reader, and do the same to put it back. Mary noted that he had to bend over and reach up under the engine to scan it, noting that it might be better to read the bar code as the frame came to the station and before the engine was mounted. When George had completed the entire process, Bill noted the time at 6:05 minutes, 35 seconds over the takt time of 5:30. George repeated the simulation two more times, coming in at 6:10 and 5:54.

After the simulations, the team filled out a Standard Work Combination sheet (see Exhibit 8.5), documenting the sequence of steps that George had followed, and the average times from Dave's Time Observation form, and then they updated their Yamazumi chart which showed the total work content at the station when all of the individual steps were added together. They met as a team and reviewed each of Mary and Pete's observations, considering ideas to address each of them. Where appropriate, they also updated their Process-at-a-Glance sheets.

"Hey Ken, we're running out of wall space. The Process-at-a-Glance sheets are taking up so much room," Mary noted.

Date:	Work Sequence: of				Demand:	Manual ——— Machine ...... Walking
Part Name:					Takt Time:	
STEP #	WORK CONTENT DESCRIPTION	TIME			Work Content Graph (.1 sec/div.)	
		MANUAL	AUTO	WALK	5 10 15 20 25 30 35 40 45 50 55 60 65 70 75 80 85 90	

Exhibit 8.5 Standard work combination sheet.

Ken called the whole team over. "Team, as we get more confident that each of our station designs is sufficiently detailed, we can begin to archive our Process-at-a-Glance sheets and free up wall space. Our walls are starting to get a little too cluttered, and we always should only keep things on the wall that add value and those we'll need to continue to refer to and/or update. Mary and I will create a simple storage area for our documentation. The only things we need to keep up all the way through the project are our Yamazumi and Standard Work Combination sheets." The team members nodded their understanding, and got back to work.

They made the changes in work sequence and method and improved the mockup stations based on their observations during the simulations. In less than an hour and a half, they managed to move the point-of-use material shelf closer and changed the order of how they stored the material. Then Ken asked them to simulate again after the changes were made. This time George was able to finish below 5:15 minutes each time, a significant improvement. They went on to conduct a similar exercise on two other station mock-ups. On the second station, they conducted two simulations, with team members observing and recording, and made improvements on the mock-up and work sequence after each simulation, updating their documents in pencil each time. Ultimately, they got below takt time.

At the last station, they went through four different simulation exercises, with Mary taking copious notes on ideas for improvements. The first round on the third station came in at 5:25, which was below takt time, but also showed that there were many opportunities to reduce the process cycle time and improve ergonomics. They also found that the fixture they had built from foam to mistake proof or 'poke yoke' the process did not work that well, and Dave came up with an idea to improve it and make it almost foolproof. By the end of the fourth simulation, they had gotten the process time down to 2:30, well below takt time. This provided an opportunity to simplify the line further by possibly reducing the need for another person as well as a station, if this process could be combined with another station and still keep the total work content within takt time.

Next, Ken had them conduct a simulation of all three stations at the same time, and had them move the units at the end of each cycle to the next station, to ensure there were no unforeseen problems with moving. In this case, Bill and Dave and George all simulated the work, following the latest version of the Standard Work Combination Sheet. Mary and Ken observed this simulation, since Pete had to leave.

All around them, the other three teams went through the same exercises. Each team built scale mockups from whatever material they felt was best, and adjusted them as needed. At that point they updated the Process-at-a-Glance sheets. Next, they conducted simulations, process timing and observation. Finally, they implemented all improvements agreed upon, and updated the documentation once again. They continued to explore all of their ideas to eliminate waste and made sure that each station was below takt time.

When Dave's team had finished with the first three stations, they moved on to the fourth. The fourth station was complicated, with a lot of fixtures, different materials, and required many small steps. Ken advised them to stay with the process and not be overwhelmed. Once they were satisfied with what they had at this station, they could then move on and continue down the line. Bit by bit, day by day the floor had more and more station mockups, the process coming together like a large puzzle.

This exercise continued for four weeks. The teams followed the same model over and over as they moved through the Process-at-a-Glance sheets, experimenting with different ways of making the mockups, and repeatedly simulating and observing, removing waste through various improvements, and updating their documentation. In the end, they'd represented the full process in a 1/5 scale mockup, incorporating all of the ideas they'd identified to date. The end result can be seen in Exhibit 8.6

George was huddled to the side of the room on a break with Dave and the three hourly moonshiners, Bill, Norm, and Byron. He said to the others, "Can you believe that we've only been at this for ten weeks? We have the entire new process laid out and we're still about ten months from launch. I've never seen anything like it. In the past, we didn't even have all of the documentation or processes completed at launch. I can see now what Ken meant when he told us we had a great opportunity for a vertical startup. This is going to be great."

Bill added, "I wasn't sure about building the mockups with foam and cardboard. It seemed pretty childish. But we've gotten creative at building some realistic mockups."

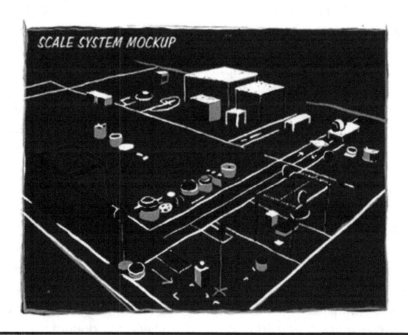

Exhibit 8.6 Scale system mock-up.

Norm added, "I know what you mean, Bill. When I saw you gluing that first balsa wood arm to the side of the foam, you did look like you were a twelve year old again. I had the same reaction to the simulations. I wasn't sure it would prove anything without being in real production, but we were able to simulate what will really happen pretty well. I think if we had the real product and equipment here right now, we wouldn't be that far off."

"I agree, Norm," Dave added. "I'm surprised at the level of documentation we have at this point, so far from launch. I've been preaching about standard work for years, but this takes it to a whole new level. We're still in simulation mode, and doing kaizen on mockups, yet we're developing really solid and realistic documentation along the way and keeping it updated. A lot of teams have a hard time keeping standard work updated in the real production world, but the discipline in 3P is really high."

Ken asked the whole team to work on more simulations for the feeder areas and the main line, to ensure they had their process times and movement of the product worked properly. They looked for high-level wastes in the general sequencing of the line. After so many simulations on individual stations, they thought they had eliminated all of the waste, but were surprised to find more opportunities for improvement. They observed one problem at the test station. They'd been counting on getting yields up to 98%, from Trail Gripper's current 91%. Even at 98%, they knew they would still need the ability to pull units off the main line to a rework station, a non-value-add step, but a necessary one. Getting the units that failed test off the line was not a problem. The units could simply be rolled over to an area that would be designated for rework. Their issue was the timing and methods to get a reworked unit back into flow, without disrupting the cadence and performance of the line. They solved this problem by establishing the practice of inserting the reworked unit at the station immediately after test. The next time another unit failed test and was pulled off. This would maintain the flow of the rest of the line. By then, the team had learned just about everything they could with the 1/5 scale mock-up. Although Ken and Dave knew there were more opportunities—there were always more opportunities—it was time to move on.

Pete was sitting at his desk when his phone rang. He picked it up to hear the receptionist announce that Mr. Sawyer had arrived for his scheduled visit and was waiting for him in the lobby. Pete walked down the hall to meet him.

"Pete," said Steve, "how is my protégé doing?"

"I'm doing well, Steve," Pete replied, shaking his hand.

"Seems like you've been doing better than well. I've been real proud of how you've turned things around. Tell me a little more about this 3P process I've been hearing about."

"I'll do you one better," Pete responded. "I'll show you. Come meet my team."

Steve Sawyer came in to the team room with Pete. The timing for Steve's visit couldn't have been better, since they'd just finished the scale mockup exercises.

They had created a model of the entire process and the whole line could be seen at a glance.

They walked Steve through the room, showing him the process from the start. Each team member took turns talking through an exhibit or visual display. Steve asked a number of questions When they left the room to see the scale model of the line, Steve had only one word for what he saw. "Wow!"

Steve could already see that this method could be a game changer for all major programs in the company. "How long did you say you've been at this?"

"Ten weeks," said Pete.

Steve replied with a simple "Unbelievable."

The team walked Steve through all of the process elements, looking at the related Yamazumi charts and Standard Work Combination sheets.

At the end of the day, Steve thanked the team, gave them a positive summary of what he had seen, and concluded by reviewing verbally with them their accomplishments for the company and their customers the past two years. He explained to them that the annual corporate employee survey showed their employee engagement levels as most improved in the company by far, near the top of the list overall. He told them that the engagement scores were perhaps the most meaningful to him as he had learned, long ago, how much difference a motivated, positive, energized and empowered workforce could make. He promised that he'd be back, and because he was so interested in this new process, he'd be bringing a number of other company leaders with him.

"What's next, coach?" asked George Hall. He'd come a long way in the past year.

Ken replied, "Glad you asked, George. We're going to go through the exact same exercise as we did with the scale mockups, but we're going to build full-size mockups instead."

Sylvia said what everyone else was thinking. "Ken, I don't understand why we want to do that. We've developed a great process already, and we're having a hard time seeing any more wastes."

"I can understand why you feel that way. Your team made so many improvements to your section of the process, it was hard to believe. You should be proud of what you worked so hard to create, but I've always found that waste becomes clearer the closer we get to touching the actual tree."

"Remember when I talked to you about looking at a landscape from a distance? As you got closer to it, the mountain details come into view. When you got even closer, you could start to make out details of the forest. Even closer, and you could start to make out individual trees. And when you got right up to the tree, you could see all of its detail and you could touch and feel it. When we were doing the 2D line layout, we were working at a very high level, but we kept making changes, over and over, as problems jumped out at us. We might have thought that we'd created the ideal flow process then, since it was so much better than what's currently in practice at Trail Gripper. Then we did the Process-at-a-Glance exercise, and you were able to get into more detail and see more problems and opportunities. The 7-ways work helped even more with this, and when we made those bench top mockups, the wastes became even clearer. Things you couldn't see in a sketch, suddenly jumped out, and those bench-top mockups looked pretty good, didn't they? But when you took them and moved them to 1/5 scale, even more wastes became clear, which hadn't been evident at bench top size."

"As we keep going forward, we'll help you see more and more opportunities for improvement. When we get to a full size mockup, we'll expose more waste. When we take those full scale mockups and build prototype equipment, we'll expose even more. Our objective is to eliminate as much waste as possible before we hand the process over to the production operators. But no matter how much of this we do, until we have the actual production equipment, and follow the real standard work process with real production parts, and do this at the target takt rate, we won't be able to see all of the opportunities for improvement in our process. So, we'll do simulations, observations, and kaizen on full-scale mockups, on prototype equipment, on production equipment, and on the real process at full takt rate. All of this will be done before we start up the line in the factory. We'll have a successful launch. But when more and more operators use the real process over and over, they'll still uncover problems. We'll never eliminate all of them, even after years. After all, the real objective of Lean is to continuously improve—to expose problems, solve them, improve the process, and repeat all over again."

John Lee understood the explanation, but was still curious.

"Ken, this all makes sense, but why do we have to go through all of these stages? Why not just get to the final production state and do a ton of simulations and observations, where the wastes will be so obvious and clear?"

"Why do you think we wouldn't do this, John?"

John thought for a moment. "I think we want to go through all of the stages is that if we tried to get to the finish state in one go, we might not find the tree we want. If we start at a high level, we get a full view of the landscape, and it helps us to find the right mountain. When we get closer, we can find the right section of forest because we can see it all on the mountain, and eventually we'll get to the right tree, the one we want. If we don't start from a distance with a wide view, we might miss the ideal end state."

"That's exactly right, John. Very good. By starting from a distance or a height far above the target and moving closer and closer, we can home in to exactly the point we want, and end up at the ideal state. We know that even when we find the ideal we can always keep improving it. Nevertheless, we want to start up our new process on time, at the target cost and quality, and with the ability to deliver high levels of service to our customers. We'll get even better from there. We'll get there by working through these phases with minimal focused details but a wide range of peripheral possibilities, and then narrowing our focus bit by bit. Does everyone get this?"

"Yes, we do!" the team shouted back.

George Hall, the AME, spoke up. "Besides, it's a lot cheaper to make process changes on mock-ups then on production equipment.

Brenda added, "And it's a lot less expensive to make product design changes on the 'drawing board' than when we're in production."

Everyone nodded in agreement. Ken said, "Let me tell you about the 'rule of tens.' It reinforces what you're both saying.

RULE OF TENS

There is a common rule of thumb: that every $1 correction on the proverbial "drawing board" can cost as much as $1,000 if left uncorrected and discovered by the customer. The cost of correcting a product or service failure increases incrementally—as much as tenfold—as it proceeds through each major stage of development, from concept to the customer. So a $1 correction while still in the design stage will cost $10 to correct at the prototype stage. It will cost $100 to correct if it's found in production, and $1,000 if it's allowed to escape to the customer. Although the specific dollar amounts may vary based on the particular product or service, the point remains: a poor design can add significant cost to any business.

Ken continued, "So, when we move on to full-scale mockups, we need to be careful about what we decide to work on and what we don't. There are always limitations to the process. Sometimes, we don't have enough room to lay out a full-scale factory. We have a lot of room to experiment here, but even we don't have enough room to stage the entire process. We can mock up a group of connected stations, and go through the simulations, observations, and improvements until we identify all the wastes we are likely to find. Then we'll move them over to storage, and start a new group of full size mock-ups.

"We do want to try and mock up as many of the stations as we can, but where we have limitations there are some things we should consider. Anything that has strong direct impact on the external customer should always get a full size mock-up. Any station that has a lot of changes from the original station should be mocked up. Any station that is complex and large, and any station where we're still unsure about what we ended up with on the 1/5 scale mockup—we'll do all of these for sure. We do have a lot of space and we do have time, so let's start off and try to get to all of our stations."

Ken split the teams into the same groups, with the exception of Scott Green, the material controller, and his team. They were pulled out of the scale mockup exercises about a third of the way through. They had started to work on the information and material flow processes, and picked up on those where they'd left off. Everyone went right to work.

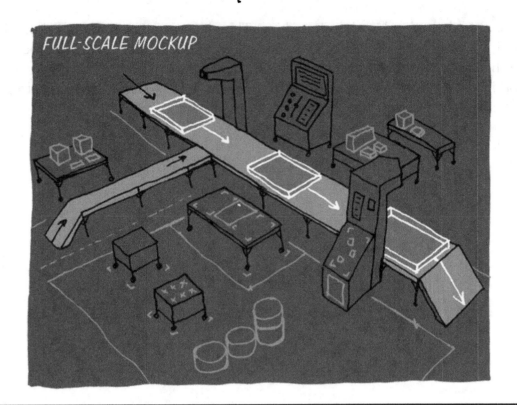

Exhibit 8.7   Full-scale mock-up.

Right out of the gate, the team had to start thinking differently about how to mock-up larger parts and assemblies, along with furniture, carts, equipment, fixtures, and tools. A lot more materials would certainly be needed. Before long, they found envisioning and building mock-ups to be almost second nature. Over the next twelve weeks, they went through a repeating series of mock-up builds, simulations, observations, time studies, continuous improvement, and documentation updates. Through it all, they were able to see more problems and opportunities for improvement than they could when they were looking at scale models, and their process design continued to get better and better and better (Exhibit 8.7)

Steve Sawyer had been so impressed with what he'd seen, he was back again with a group of people representing various Enterride businesses. Pete and Dave acted as proud hosts. The team walked their guests through the whole process, from start to present, using all of their visual displays and tools. The scale mockup area was left intact because it provided a holistic system view, and many of their full-scale mockups were out on the floor. The team room still had much of the documentation up on the walls. The Yamazumi charts and Standard Work Combination Sheets had been moved out into the full-size mockup areas on simple rolling display boards so that they were easier and quicker to refer to and update.

The cross-division team from Enterride took copious notes. It was clear from the feedback sessions that the 3P process was about to start to spread far and wide throughout the corporation. A plant manager summed it up best before

Steve closed out the day's benchmarking. The plant manager commented that in all of the years he had been working, he had never seen such a simple process that had so much obvious power to advance their overall process capability to world-class levels in one go, and he, for one was going to learn more about it, and take the process to his people.

Chapter 9

The Other Flows

About a third of the way through the 1/5 scale mock-up process, Ken had pulled out a team to begin work on the information and material flow processes. These would include the manner by which the line would be scheduled, how the sub-assembly or feeder lines would know what to make and when to make it, how purchased items would be replenished from outside vendors, and how the purchased parts and materials would be delivered to the various points in the production line. These processes were as important as the product flow process that the rest of the team continued to work on.

Scott Green was leading the team, which also included Betty King, Linda Campbell, Bill Stark, and Johnny Cox. Johnny was an hourly associate who had come out of the assembly department. He'd spent years as a stockroom employee and a material handler, delivering materials to the production line. Johnny had also worked in several production areas. Armed with this varied experience and a "can-do" personality, he would prove to be a valuable member of the team.

Although this particular team was no longer working on scale mock-ups, they continued to work in the same main area where the scale models were being built, just off to the side of the war room. They had rolled out two 4 × 8 foot whiteboards to visually display all their documents.

Ken began, "What we'll do first is to develop an information and material flow diagram. We'll start with the customer who places an order for a mix of Trail Gripper products with different options and in different colors. This begins the information flow. Of course the customer also receives the completed order, which marks the end of the product flow."

Over the next several hours, they went on to show how orders from all customers would be received, entered into their system, and scheduled on the production line. They then worked on the processes of knowing when to purchase items from vendors, scheduling the vendors, as well as the information flow showing that the materials were received into the facility and ultimately moved into the stockroom or warehouse. They'd all agreed that once the materials were in the warehouse their movement to the production line would be triggered by simple visual and manual information methods, also called "kanban." There was no need for the computer-based scheduling system to do this, which often adds complexity. Simple and visual is always a better approach.

They also determined how material would physically flow from suppliers to the factory. The team identified that they would need a warehouse to store the material when it was received from the vendors, and that it would serve as a "supermarket" for purchased materials. This would ensure that they always had what they needed, but not too much—just like a supermarket. The materials would then be moved and stored line-side at "point-of-use" at various locations in the production process itself. They posted the drawing depicting the information and material flow on the wall, as seen in Exhibit 9.1.

The team agreed to use the existing Trail Rider Enterprise Resource Planning (ERP) system for the information flow processes, as it had only recently been implemented at the site three years ago and it was becoming a corporate standard throughout Enterride. They would load the Trail Gripper Bill of Materials (BOM), which lists all the materials needed to build the product, and all other required information into the ERP system. This information would have to be revised as the 3P team did its work. For example, any changes in the product design would have to be reflected in the BOM.

The current manufacturer of the Trail Gripper did have a front-end order "configurator" as part of its order entry system, which prompted the person

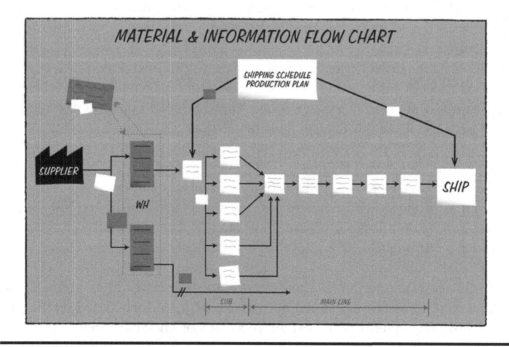

Exhibit 9.1 Information and material flow map.

entering the order to input the various features and options, or configuration, for each vehicle. An earlier team consisting of members from the Memphis operation's Customer Service, Product Design, and Information Technology (IT) departments had reviewed their systems in preparation for the transfer and concluded that it was an excellent system that could be integrated with the existing Trail Rider ERP system. Consequently, Pete had asked a separate IT team to make it happen.

The team asked Betty King to work with the IT team as a representative from Trail Gripper. She'd help ensure that the system was correctly set up so that the Trail Gripper line would be properly scheduled by the master scheduler. It is the master scheduler who has responsibility for scheduling all the production lines in the facility. Control of the production flow within the line would be accomplished using manual kanban cards and manual scheduling boards. Simple visual signals would be used to trigger replenishment of sub-assemblies from the feeder lines. The ERP system wouldn't be used for these, as it would add a lot of unnecessary complexity. It's a lot easier to update a manual board to reflect last-minute changes on the line than to have the master scheduler do so with the ERP system, which is only run once a day.

Now that the team had developed a general concept, as depicted on the map in Exhibit 9.1, Ken started to walk the team through what they'd be doing next. He said that the bulk of their work would be to develop a Plan for Every Part (PFEP). They would develop the details for each part, showing how it's shipped from the supplier, how it's transported and received, as well as how it flows within the warehouse and factory. They would also develop standard work for material handlers, also called "water spiders." The standard work would display the movement of the water spiders through the facility to replenish materials line-side at point-of-use. The timing of the replenishment would also be worked out so that the water spiders had a standard routine.

The team gathered as much of the material data as they could get from Trail Gripper, printed it out, and posted it on their boards. They had the Bill of Materials (BOM) for each model of Trail Gripper, the customer demand data for the past two years by model, all the ERP planning parameters for each part, and all the supplier information. Because this was a transfer of an existing product, rather than an entirely new product, a significant amount of data was available. Using the data from the current manufacturer would give them a good starting point. Ken pointed out this advantage to the group and added,

> "Even with a new product, there's typically an existing supplier base that a company has been working with and a lot of resident knowledge within a company's supply chain professionals. Because of that, a lot of the data needed can still be quickly obtained if you get the right people involved."

Without strongly criticizing the current Trail Gripper team, Ken reminded the team members that they should use all the information as a valuable baseline, but that their goal was to develop a world-class material flow process right out of the gate. They shouldn't assume that the baseline data was the "final word," but needed to make adjustments as necessary.

He reminded them that, like the work they had already been involved in to develop the production flows, the way they'd get to world-class status for inventory management was to look at the baseline processes and planning data, identify all the wastes, and find ways to reduce them. They would work hard to flow each part faster through the process, from the supplier's shipping dock to shipping the finished product. They needed to shift from few movements of large batches of material to many frequent movements of much smaller batches of material. The team needed to reduce the current quality defect rates, the number of times a part was handled, the distance parts traveled, and the time people spent looking for and picking and placing parts. There may be other wasteful activities that they uncover as well.

Linda raised her hand. Ken spotted her as he was turning to look at the board. "Yes, Linda."

"I understand why we'd want to flow the parts faster through our factory, and I think we can do that, but you told us we have to really focus on reducing costs as well. If we reduce the shipment sizes and increase deliveries, costs are going to skyrocket. I studied the Trail Gripper processes. They did a good job ensuring that they had full truckloads on many of their incoming shipments. If we back off from this, costs will go up … a lot."

"Linda, that's a great point. You're absolutely right. It's very possible, even probable, that if we just work to increase the speed of flow of incoming material, then costs will go up. But we can have our cake and eat it too. What do you think we could do differently to increase the speed of flow of incoming material, while not just keeping costs from going up, but lowering them?"

"I'm not sure. Maybe we could create a route for our local and regional suppliers and go from one to another picking up the parts for delivery."

"Great! Maybe we could also drop off returnable containers when we pick up parts at each location on the route. This would reduce the cost of packaging, and the containers could serve as visual signals or kanban to replenish. This would simplify the ordering process."

Scott added, "I just read a Lean material flow book. It talked a lot about cross-docking for incoming and even outbound material. Cross-docking brings material from a group of suppliers far away from us, but relatively close to each other, into a central location and then groups the parts together and sends them in a full truck directly to us. The book also mentioned that we can ask suppliers to not just ship a given part in large batches in order to fill their trucks, but to 'kit' a group of parts together, so that the lot sizes of each are smaller, but the grouping of many lots of different parts into a kit would increase the cubic volume of the deliveries. This might work with vendors who are supplying multiple parts to us. I think we can do what you're asking,"

"You're right, Scott. There are lots of great techniques that we can employ, and I'll help all of you find your way to many of them."

"Why would we limit our improvement efforts at the shipping dock of the Trail Gripper suppliers? They have long cycle times internally, and this is going to make our planning cycles and lead times a lot longer," Betty noted.

"Another great question. You're right. There will be some natural limitations on what we can achieve if we don't work to reduce our suppliers' cycle times within their facilities, especially as our suppliers' cycle times are impacted by their suppliers' capabilities, and so on. The degree of difficulty and the scope of the project would increase exponentially if we also took this on now. We need to get our own house in order before we try to help our suppliers 'Lean out' their processes within their four walls. We'll get there in time. In the meantime, there are techniques that we can effectively apply to help our immediate situation. We can work on improving the information flows and logistics processes between ourselves and our suppliers. We can also move from a forecast-based push system to a pull replenishment system. This will help our suppliers by reducing the impact of the inaccuracy of *our* forecast, which we all know exists, on their scheduling systems. There should be a lot fewer last-minute schedule changes."

"We have so many parts," Bill said. "It all seems like so much work."

"It would be, but we're not going to treat all parts in the same way. We're going to segment our parts in a number of ways, for example, by 'A,' 'B,' and 'C' class. You've all heard of this, grouping 80 of the annual buy into an 'A' category, 15 of the annual buy into a 'B' category, and 5% into a 'C' category?" Everyone nodded. "Just in case, let me explain. Experience has shown that approximately 20% of the total number of different parts we purchase make up about 80% of the total amount we spend on all purchased parts. If we just focus our attention on the 20% that are the 'A' items, we can make our task significantly easier and still make a big impact.

"We'll treat each of these very differently, and the truth is that we'll want really fast flow on 'A' items, medium speed on 'B' items, and a decent but relatively slow speed on 'C' items. Remember that only about 20% of the parts will be 'A' items, so we will focus our intense work on these."

Ken told them that ultimately their goal was to get the right part to the right place, at the right time, in the right quantity, with the right quality, and presented the right way.

"Oh. That's all you want," Johnny chuckled, causing everyone to break into laughter. But it did seem like a lot to ask for.

Ken gently responded, "Let's keep a few things in mind. First, it's a tall order, but it's possible to achieve this for all the components. Other companies have done just that. Second, like everything else we're doing on this project, this is really our ultimate goal, and we want to try to get as close as we can to the ideal state. The reality is that it'll take time, literally many years. We want to make big improvements from our current state. We want to get to 20 inventory turns, which is a big number but not unrealistic. We want to do this with a high degree of reliability so that the users have the parts they need available when they need them. In fact, if we have to sub-optimize, it will be on speed of flow; it won't be on the quality of the parts or their availability. I think you'll make some great improvements; you're very capable and have great attitudes. Just believe that you can, and you will. Team, I can see some concern in your faces about some of the things I've been saying. That's normal; it really is. We're suggesting goals that none of you have ever thought possible in the past. I remember some of these same looks when we started to do the Process-at-a-Glance sheets, yet look at all the great improvements you were already able to make. I'm asking you as we go into this to have faith that there are similar improvement opportunities in the information and material flows."

Scott asked Ken, "I have been so much into Lean the past two years since Dave arrived, and have read many books on it, particularly on material flow methods, that when you talk about 'techniques,' are you suggesting that the 3P process will teach us these as well?"

"Not exactly. We'll certainly make use of several key 3P concepts. For example, we'll consider different alternatives, and we'll repeatedly simulate the information and material flows. You've already started this work during our discussion on different ways to keep transportation costs low while meeting the objective of more frequent deliveries of smaller item quantities. When I reference techniques here, I'm really saying that there are many concepts in the Lean 'toolkit' to help us create great material and information flows. A pull material replenishment system using kanbans is just one example. I'll help coach you on many of the principles and practices of Lean that can help you overcome the barriers we'll face. As for 3P, just keep thinking like 12-year-olds, trust the process, and you'll get there. Are we ready to get on with it?"

A chorus of "Yes!", "Absolutely!", and "Let's do it!" came from the small group, and so they were all off on this round of adventure

Ken asked the team to segment their materials into A, B, and C classifications, set targets for total inventories, and post the targets up on their whiteboards. Ken told them that if they targeted A items between 5 and 10 days, B items between 20 and 30 days, and C items between 30 and 40 days, they would get close to 20 turns for total inventory. They agreed they were going to talk in terms of "days on hand" for their metrics now, and also agreed they would target A items to a week between the stockroom and line-side point-of-use storage, B items to 20 days, and C items to 30 days. And they figured this would give them incentive

```
Plan for Every Part (PFEP) Questions

What is basic information about the part?
•   Part #, description, daily usage, value, weight, supplier, supplier
    location, lead time, supplier performance
How is the part purchased?
•   Order frequency & quantity, method (discreet or blanket PO)
How is is received?
•   Container type, quantity, size & weight, shipping method/carrier
Where is it to be stored?
•   Location(s)
How is it to be delivered to point-of-use?
•   Usage locations, hourly usage, frequency of delivery
```

Exhibit 9.2   Plan for Every Part (PFEP): Questions.

to break the stretch target goal, with some room for a few slips in areas where they couldn't implement sufficient process improvements in time for the launch.

Once they were done, Ken split them into two teams. He posted a series of questions on each team's board (Exhibit 9.2). He instructed the teams to use the questions and answers to help them to develop a plan for every part.

Ken put up a form (Exhibit 9.3) that they would fill out to document the full plan for each part. The plan would include ordering methodologies, transportation methods, packaging type and quantity, internal move and storage plans, and Lean elements.

Ken then provided the team with an example (Exhibit 9.4). Ultimately, they'd fill out one form for each part.

Ken described each of the data items in each section. Most were straightforward, and he would coach them through how to think about the best plan for

Basic Information:		
Part #: _____ Description: _____		
Value: _____ Weight: _____ Daily Usage: _____		
Supplier: _____ Location: _____		
Lead Time: _____		
Order Method: _____ Frequency: _____ Quantity: _____		
Shipping Method: _____ Carrier: _____		
Container Type: _____ Quantity: _____ Weight: _____		
Size: _____		
Storage Location: _____ Usage Location(s): _____ Hourly Usage: _____		
Frequency of Delivery: _____ Responsibility: _____		

Exhibit 9.3   Plan for Every Part (PFEP): Form

Basic Information:	
Part #: <u>13598</u>	Description: <u>Valve</u>
Value: <u>$5.99</u>	Weight: <u>1.2 lb.</u>
Daily Usage: <u>80</u>	
Supplier: <u>ValveTec</u>	Location: <u>Philadelphia, PA</u>
Lead Time: <u>1 week</u>	
Order Method: <u>Kanban, Blanket PO</u>	Frequency: <u>Weekly</u> Quantity: <u>400</u>
Shipping Method: <u>Truck</u>	Carrier: <u>UPS</u>
Container Type: <u>Box</u>	Quantity: <u>20</u> Weight: <u>24 lb.</u>
Size: <u>12″ × 10″ × 8″</u>	
Storage Location: <u>WH D2-A</u>	Usage Location(s): <u>POU-1A</u>
Hourly Usage: <u>5.7</u>	
Frequency of Delivery: <u>Every 2 hours</u>	Responsibility: <u>Water Spider #1</u>

Exhibit 9.4 Plan for Every Part (PFEP): Example

each part. He said that once they got the hang of it on a few of them, many of the patterns would be similar. He divided a list of parts from the Bill of Materials between the two teams. They focused on the A items first, then the B items, and then the C items. Over a four-week period, they'd captured an initial plan for each of the parts, and using this, they were ready to move on to the next stage where they'd develop full standard work, do simulations, and drive continuous improvement.

On Monday morning, both teams reconvened. They'd been working diligently on the first draft of a plan for all their parts. Ken said,

> "We now need to discuss the details of the stockroom or 'supermarket.' We need to make decisions on its location, how it will be organized, and other important details."

Ken pulled out a chart (Table 9.1) and explained that they should work to ensure that their stockroom design met all the criteria on it. Eventually, they would simulate their proposed stockroom design and test against the criteria listed on the checklist. The team read through all the criteria together, and Ken asked if there were any that were unclear.

> Betty said, "I'd never really thought about all of these together. They all make sense, but I'm not sure what Number 11 means. What is 'one-way kitting'?"
> Ken explained, "What we want to do is to arrange the parts in a manner so that the people who'll retrieve the parts in the stockroom can flow in one direction continuously up one aisle and down another and so on, and not

Table 9.1 Supermarket Evaluation Checklist

	Criteria	Yes	No
1.	Is First-in-First-Out (FIFO) maintained?		
2.	Are products and quantities visually identifiable?		
3.	Is damage prevented during storage, loading, unloading?		
4.	Does picking of one item take ≤3 seconds?		
5.	Is "1-part – 1-location" being practiced?		
6.	Are locations on the "pick list" or kanban card?		
7.	Is an appropriate method of picking being used (e.g., cart)?		
8.	Is the re-order point defined and clear?		
9.	Is the re-order quantity defined and clear?		
10.	Are the items stored to allow for "one-way kitting"?		
11.	Are forklifts not utilized?		
12.	Are physical inventory counts performed periodically?		
13.	Are defective or unnecessary materials segregated?		
14.	Is the storage height limited to 5.5 feet or less?		
15.	Are storage locations "right-sized" for the parts/containers?		
16.	Are like parts stored together?		
17.	Are there any safety concerns? Are heavy parts stored at proper level?		

have to back up to get a part they already walked past. This is intended to reduce the wastes of motion and transportation, and speed up the process of retrieving parts in the stockroom. Does this make sense?"

Betty nodded.

Ken encouraged the team to begin developing a routing and flow map for their PFEPs, in preparation for simulations. He told them it was similar to the 2D production flow diagram they did earlier in preparation for Process-at-a-Glance sheets. The team took a layout of the plant from the most recent scale mock-up that they had. Ken asked them to determine how often each part would be replenished on the line based on the takt time. They made a list of the estimated replenishment frequencies for each item, along with an estimated time to place the parts and pick up a replenishment signal in whatever form they had. They then accumulated all these times. Ken asked them to plan a route through all material areas of the plant, then measured the travel distance and estimated the time to travel all of it with a material handling "tugger" at nominal speed. They added this to the total material drop-off times, and it added up to 1,142 minutes.

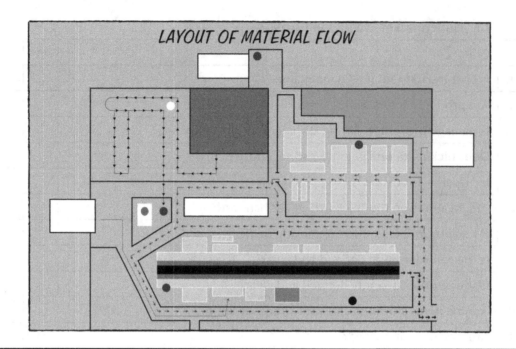

Exhibit 9.5   Material flow plan with water spider routes.

Ken sat with the team around a table, and they estimated the effective working time per shift for a material handling person, or "water spider" as he called them. They estimated that each water spider would be available 420 minutes per shift. When they divided the total time by the time per person (1,142 divided by 420), it came out to 2.72. Ken told them they'd build a material flow plan around three water spiders per shift. On a 2D drawing, they plotted three separate routes using three different colors on the map, showing the direction of flow for each (Exhibit 9.5).

They made a list of all the components for each route that would need replenishment, and sorted the list into the order along the route they would be replenished. They then developed a standard work route map that listed the drop-off/pick-up tasks at each location and the moves between drop-off/pick-up areas. Finally, they added estimated times for each task and each move. When they were done, all three routes came in under the 420 minutes of available time.

The teams made mock-up locations for all parts, using cardboard markings, and placed them at their appropriate locations line-side on the 1/5 scale mock-up layout. They also laid out the stockroom, using sheets of cardboard to simulate rows of shelves and penciled in where parts would be stored throughout the stockroom. The team then spent the next four weeks doing simulations of receiving parts into the stockroom based on the PFEP plans and moving them through the water spider routes. Observers and timers documented the process while they simulated the flows. During this period, they continuously updated locations, the PFEP plans, and the sequence of steps and timing, trying to reduce waste, simplify the flows, and reduce the material processing times.

When they finally felt that they had a good plan, and while other team members were developing equipment specifications, they worked with the maintenance department to build and assemble shelves and delivery carts as per the PFEP plans and located them in the new plant area that was under development. The main effort of the information and material flow team was now complete. The team was confident that what it had developed would work. Of course, everything is subject to further improvement, but they didn't expect anything significant.

After the rest of the Trail Gripper team completed the full-scale mock-up simulations, they developed specification sheets for all the equipment, furniture, fixtures, and tools they'd developed. The manufacturing and product design engineers in the group took the lead in this effort, creating technical specifications for the more complicated pieces of equipment. Everyone else focused on the less-complicated furniture, fixtures, and tools. They specified sizes, weights, load requirements, quantities, preferred materials to use, etc. Ken spent all his time going from person to person, coaching them through the tasks. The team continued with these activities for the next four weeks. They then sat as a team and decided which equipment they could order from a catalog, which they'd try to build internally with their three moonshiners and their maintenance team, and which they would contract out to the three external engineering firms that they'd use. They ended up deciding to build about 35% of the equipment internally.

The team spent the next 14 weeks building the equipment, ordering the catalog items, and getting quotes from the external firms and authorizing them to build. Most of the equipment was relatively simple and wouldn't take a long time to procure. However, there were 12 pieces of equipment that required outside engineering. The team had the three engineering firms provide quotes for these, and then begin to build them. This equipment was more complicated and required a 20-week lead time.

Everything was underway. Several team members were assigned to help monitor the design, fabrication, and assembly at the equipment suppliers. Several were assigned to help set up the material racks and warehouse and prepare bins, kanban cards, etc. Others were assigned to plan the pilot-testing and training phases, while still others performed additional simulations, but this time using the actual production equipment as it became available. They repeated the same standard work simulations that they'd done on the full-size mock-ups, with observers, a timekeeper, and a rotation of team members simulating the work. They found some other waste opportunities, and the moonshiners made some small adjustments. These small errors were primarily in the specification, design, or manufacture of the equipment itself, along with some errors they hadn't anticipated from the mock-up in the way the equipment, fixtures, and tools would work.

The maintenance team was very busy during this period and made the small corrections and adjustments. All in all, the work went quickly. There were a few mid-sized problems and fewer large problems. The team was strikingly close to being "production ready" with everything.

❋　❋　❋

Dave turned the wheel of his Ford F150 truck tight as he pulled into the parking lot. He walked into the gymnasium and spotted his wife in the bleachers first, and then saw the teams coming out for warm-ups. He had made it before the game started. His wife moved over and he slid in beside her. As he watched the warm-ups, his thoughts went back to everything he'd learned about the 3P process. He thought back to all his years of playing different sports as a boy, and the years playing Division II football. He remembered how much they practiced over and over, just like his daughter, to get ready for the next big game. He realized that one of the big differences between what they were doing now, versus the old days of driving major change projects, was the amount of practice and preparation they were doing. He realized that there isn't a team that can compete in their league without dedicating themselves to a lot of practice.

And they'd been practicing. Practicing, looking for mistakes, correcting the play, and practicing again, over and over, one different sequence of plays at a time. He also realized that the pre-launch training scheduled for their hourly operators would be practice for them. They would be practicing their craft, using standard work, *before* the launch date. They would be learning ahead of time by trying, by doing, over and over, the same plays in the same way, over and over, just like learning a new sport.

It dawned on him that a significant amount of up-front practice and improvement was a key part to any successful change effort.

❋　❋　❋

Chapter 10

The Vertical Start-Up

They were just ten weeks away from the Trail Gripper launch. Although they were coming down to the wire, no one was panicking—a different experience for a project of this magnitude. Pete was confident that they were going to achieve vertical start-up, as targeted. They just had a few more phases to go before full production launch: production testing the equipment and processes, and taking the broader team through training and start-up activities.

Ken and Dave toured the new addition to the factory, a gleaming sea of blue-gray concrete with clean white lights. All the utilities were in place, with the utility drops spaced in equal patterns, waiting to connect to the equipment that would be part of the new line. A few members of the team were busy laying down masking tape to mark where the equipment, fixtures, and furniture would be located, and were preparing to paint all the visual management markings they had developed in the mock-up stages, from aisle-way markers, to safety zones, to kanban locations. Some of the equipment, furniture, and fixtures that had arrived were already in place on the floor, and some of the pilot testing had been completed.

"So what's next?" Dave asked Ken.

"I talked to Pete, and we want to start testing the production equipment and help train the operators. We're going to start running the line at a 7.7-minute takt time, to match the initial volume projections, and use the Standard Work Combination sheets that were developed for that rate. At some point we'll want to run the line at the 5.5-minute takt time projected out a year from launch. It'll just be a short-term test to confirm it's feasible. We'll worry about that later.

"All our pilot testing indicated that almost all of our equipment, furniture, and fixtures are ready to go, with a few final modifications. We know that the A-frame lift that we built to move the engine assembly into position is just too awkward to use in a continuous flow environment and we have Bill and Norm looking into solutions. And the device we built to tilt the frame for access to some of the mounts is just too big for the cramped quarters around it on the line. Byron is redesigning and rebuilding that one, with Norm's help.

"I want to split up the team into pairs and have them go through the inventory of equipment and make a list of items that we still need to finalize for production. I'd like you to coach the folks through it as I have other commitments this afternoon."

"Sure," Dave said.

Later that afternoon, Dave had the entire team meet in the mock-up area. He then proceeded to divide them into pairs for the next activity.

"We want you to take one last look at each piece of equipment, fixture, tool, or furniture and evaluate whether or not it's robust enough to operate in a real production environment. We've already run through numerous trials in our pilot testing. Despite a few minor issues, everything seems fully operable for the tasks required, in the way we prescribed in our early simulations and standard work development. We haven't run at full rate yet, and we still need to do that, but we're ready to get the equipment ready for production. I've split the entire list and assigned portions to each team of two. You should have that list in front of you. Any questions on any of this so far?"

"What do we check for?" Sylvia asked.

"I was just about to get to that. Use the list I gave you and note any issues that need work in the comment section. Our moonshiners will make the improvements, and, if need be, some of the regular maintenance folks will help as well. Keep in mind that we really want you to look hard with a critical eye for safety problems. You're not solely responsible for safety—remember, we checked and rechecked our simulations for safety issues and made improvements in the mock-up stages. This is just one last look. We're also going to have the Environmental Health & Safety team go over the entire line to ensure that there are no lingering safety issues.

"So, as far as safety goes, you want to look again for any exposed moving parts, pinch points, and electrical or mechanical hazards that need to be guarded. Think about ergonomics, particularly for operators of different physical sizes and strengths. Where there are repetitive tasks, make sure they're easy to do and create minimal strain. Look for any sharp edges or blunt end points that someone could fall against and get hurt. Check for any potential environmental hazards from or around the equipment.

"We also want to look at the durability, quality, and reliability of the equipment. Make sure that there are no flimsy assemblies on any of the fixtures, tools, or equipment. Make sure there's nothing loose on the equipment. Where there are things that could come loose, check that there are effective fasteners to prevent it. Make sure that any devices on the equipment that were put in place as simple mock-ups to enable testing have been upgraded to real production-ready devices.

"Be sure to identify on the list if something needs to be modified. If it can be quickly addressed, then please do so; otherwise the maintenance folks will get to it later. Pete, George, and I have worked out a schedule with maintenance and some outside contractors to hook up all utilities and do whatever checks are needed to approve each piece of equipment for use by production. While this is going on, we'll be setting up the material storage areas and moving materials into place in the stockroom and the line-side point-of-use locations. Any questions? … No? … All right; let's get started."

It took the team three full days to exhaustively audit the entire equipment inventory. They took it seriously. They knew that this was the culmination of all their hard work, and they didn't want anyone to get hurt. They wanted the equipment to work correctly and consistently, and to be robust enough to continue operating well with an effective Total Productive Maintenance (TPM) program. The TPM program would ensure that the equipment was properly maintained over time.

Most of the equipment was signed off on, "Okay as is." Over the next eight weeks, the moonshiners and maintenance teams worked to modify all remaining items. The Plant Maintenance crews were putting all the equipment, furniture, fixtures, shelves, carts, and tooling into place. George was overseeing this, much as he had similar projects over the years. He was a veteran of plant shutdowns, line changeovers, and line expansions. As they got the equipment into place, the Plant Maintenance crews and outside tradespeople were connecting and checking everything. The rest of the 3P team worked concurrently on other important tasks. They took materials that were steadily coming in from receiving and moved them into their correct locations as noted on the material flow plan the team had created (refer to Exhibit 9.5 in Chapter 9). They were building the stockroom and line-side point-of-use storage areas and then stocking them appropriately. If they were missing parts, they expedited their receipt. They moved all the workstations into place, with point-of-use parts, fixtures, and tools, based on the mock-ups they had refined over and over. With one month to go before launch, Pete was walking the new line with Ken and Dave. He had been working with the Human Resources (HR) department for the past three months, and they had created a hiring model that had most of the operators starting a month early. HR had developed a complete training regimen for each of them, which included building actual pre-production units. All the standard work was in place at the workstations. Each member of the 3P team would train multiple employees. Their goal was to get each person up to speed on their standard work to meet the takt time and achieve the necessary quality required, prior to full production.

Every day was a whirlwind of training and building of pre-production units. Some of the assemblies and full units were torn back down and then rebuilt over and over, while the 3P team worked diligently to ensure that each operator had all the teaching, coaching, and practice that he or she needed. The HR team had also created a series of classroom trainings, including indoctrination to the new plant, company information, product and customer information, teamwork, safety training, quality training, and Lean fundamentals.

Day by day, their confidence grew. The pressure was still there but the team was confident.

❋ ❋ ❋

Finally, it was the day before, and then the night before launch, and they were on time and on cost. But would they be on quality and on rate?

5:00 a.m. Launch day!!!

Pete was up before the alarm clock went off. Halfway across town, Dave was wolfing down his last piece of toast, grabbing his briefcase, and rushing out to his pickup truck. In the parking lot, Ken's headlights shone across the bottom of the security building as he turned into the front entrance. All three of them would be on the floor by 6:00 a.m. They walked out together to the new expansion area, waiting for the waves of first-shift operators to arrive, punch in, and gather in front of the new "mission control" board for an early morning pep talk. Supervisors and operators would meet at the board at the beginning of each shift to review past performance versus goals that had been set, take a look at the upcoming schedule, review any issues that came up, as well as provide general information that needed to be communicated. The ten-minute "huddle" would prove very effective in maintaining a strong sense of team over time.

The prior afternoon, the team had worked carefully to set the appropriate amount of work-in-process throughout the line to ensure a smooth start-up. The 3P team wanted to be able to test the line on launch day with the line primed. They had spread themselves throughout the line, helped the team get a Standard Work-in-Process (SWIP) unit at every station, and fill all queues to the standard quantities. On launch day they would be able to see how the line ran without having to wait until the front-end stations completed units for the downstream stations. By 6:30 a.m., the start of the shift, the entire production team and the 3P team were assembled, waiting for Pete to address the group. He walked up in front of the mission control board.

"This is a great day in the history of our plant. We're about to implement one of the most impressive product transfers our company has ever completed. We're expanding our revenues and workforce, and this should help ensure a better future for all of us. I want to thank the 3P Trail Gripper team for all their hard work the past year, and I want to thank all of you ahead of time for your efforts to make the Trail Gripper product successful with our customers going forward. Many of you have trained for the past month on your standard work; we've worked hard to reduce the number of problems we expect to have and to make the process the best it could be. However, we'll undoubtedly still have problems once we get started, and there are still many, many ways we can make the process even better. I encourage all of you to stay committed to following the standard work and making sure that we can flow well and that we make takt time. I also encourage you to help us solve any problems that arise. Together, we'll keep looking for ways to continuously improve our process over time. Let's go do this."

Most of the hourly operators were new hires and happy to have a new job. Few of them had ever received such intensive training. Spirits were high, for the new hires and the 3P Trail Gripper team alike. Hourly team leaders were assigned to sections of the line, with team leader-to-operator ratios of about 1 to 10. The 3P Trail Blazer team was spread out among the team leaders. They were committed to spending the next week observing the live process, as they had observed the simulations for all these past months. They would take all that they'd learned about waste and abnormalities, and record their findings. They were to coach and teach the team leaders and the operators on the standard work, and to help solve problems as they arose. Temporary countermeasures would be implemented to ensure that flow continued as soon as possible, with permanent countermeasures then devised to permanently eliminate each abnormality. They also would look for opportunities for improvement, although some of them felt that they had eliminated most waste before this. Ken reminded them that they'd thought the same thing many times, during scale or full-scale simulations. However, they continued to find more and more waste opportunities every time they did a new simulation. The waste started to feel endless, and he assured them that it actually was. In the real-world production environment, it would continue to expose itself continuously to those who were looking for it.

Some of the team would stay with the project after launch, filling many of the new salaried plant operating roles. A few had been promoted. Many of them would go back to their "day jobs," although all were reluctant to do so; it had been a thoroughly enjoyable experience. Launch day went well, with all stations working generally as planned. There were a lot of minor, and a few mid-sized abnormalities, but little that caused any major issues. They hit 94% takt time attainment on the first day, a remarkable accomplishment. Working just 20 minutes of overtime, they made the full rate for the day. The quality team was duly impressed by the minimal failures at test and the minor quality issues they observed on the lines themselves.

❋   ❋   ❋

Pete was elated, as was every single person on the 3P planning team. The team had gathered in the training room for a final wrap-up and celebration of the completion of the launch. Ken said he was going to go around the room, and he wanted everyone to come up with one takeaway from the Trail Gripper 3P Project.

✳ ✳ ✳

"Dave, let's start with you."

"Sure, Ken. For me, the biggest discovery is that there really is a formal standard process for 3P. I should've known it. I'd always done elements of 3P, but this pulled it all together for me."

"Very good, Dave. How about Mike, our Shop Manager? What did you learn?"

"Setting the right climate and expectations up front is critical, especially the need for open-minded behaviors. We needed to act like a 12-year-old."

"George, as the advanced manufacturing engineering lead, you've been through numerous start-ups. What did you take away from this one?"

"We always wanted to jump ahead for the answers, but we learned that we needed to have the discipline to trust and follow the process and the answers will come."

"John?"

"Mock-ups really can be made simply and easily, and for next to nothing, and they really are a powerful way to learn and improve."

"Gina?"

"It's important to do simulations over and over again on the same process. Each time, it was amazing how much new waste we found that we didn't see before."

"Mary, what do you think?"

"I was amazed at how much progress we could make in such a short period of time. I was resistant to much of what we took on because it felt overwhelming and that it would take us months to complete. But once we got into it, every day we accomplished way more than we thought we could or would."

"Sylvia, you always have such great insights. What are your thoughts?"

"We had a really good team. When I look back at it, it was obvious that positive, can-do, creative people can achieve the best final process in the end, especially if they're from diverse backgrounds."

"Bill?"

"I found the visible reminder of the do's and don'ts very important. I'm not sure why, but I always found myself falling back to the don'ts. It became clear to me that these were not the best practices, however counterintuitive it seems. We were able to move toward a lot more of the do's of process excellence because we always focused hard on them … oh, and also because Ken kept reminding us … *constantly*."

Everyone chuckled at Bill's emphasis on the word "constantly."

"Lou?"

"5S is critical, even in this work. There were times where we got a little sloppy and it slowed down our work noticeably. We were so clean and organized that it became a way of life here and we wanted to keep it that way. When it got out of hand, I was surprised how fast we noticed and did something about it."

"Johnny, how did it feel from a value provider standpoint?"

"It was really important to have senior leadership support. I've never seen as senior a person as Steve Sawyer at our plant before. Also, having Pete so actively engaged in the work made me feel a lot more confident in taking some risks. It made me believe a lot more in what we were doing."

"Gerry, you were another value provider. What did you learn?"

"I thought the manual drawings and documents that we created and corrected in pencil and marker and hung on the walls were a lot more effective than working with everything buried in a computer as we would have done in the past."

"Norm, what do you think, having been such a critical part of the process?"

"It really helped us in our work to have a well-stocked and maintained supermarket with a whole variety of supplies to help us with all the prototyping work and other activities."

"Byron, another world-class moonshiner, what did you learn?"

"I was amazed at the simplicity of the equipment that we designed and what we were able to build ourselves."

"That's a great point, Byron. Harry?"

"I thought focusing on the schedule on a daily basis, multiple times a day, and course-correcting when we needed to was important. Plus, the visual reminders were really effective. Ken always kept us attentive to the target date as well. That helped a lot."

"Brenda? You were our resident design engineering specialist. How about you?"

"I thought the Process-at-a-Glance and 7-Ways were very powerful tools, and they were easy to use, particularly the 7-Ways. I can already see how I'm going to use this in product design to develop better components and assemblies in the future that are cheaper, more reliable, and a lot more manufacturable."

"Betty, what does our commodity leader think?"

"For me, I thought that starting with a 2D drawing and moving to benchtop mock-ups, then scaled ones, and all the way to full-sized prototypes was important. The analogy of starting high up and seeing the whole landscape and moving closer and closer until you can touch the tree put it all in perspective nicely as well."

"Scott, let's write one more lesson up on the chart. We'll capture them all and redistribute to all of you."

"I felt that the big room concept worked out very well, where we had so many different cross-functional people working side by side co-located with each other through the life of the project."

"Okay," said Ken, "that was great, team, those were fantastic and truthful summaries. I'm not sure I could've captured the lessons any better. Now I want to do the same thing and go around the room, but this time I want you to give me a benefit that you think we achieved."

"Dave, why don't we start with you again?"

"Well, many of us loved the idea of a vertical start-up, but we were very skeptical based on our past histories. We proved to ourselves that it can be done. We hit the exact target date that we said we would from the start."

"Scott, you ended the last round, what do you think?"

"I was skeptical when we set a 20 inventory-turn target, yet our PFEP plan came out to 22 turns. So far it looks like we are actually achieving and sustaining that."

"Betty?"

"We cut product cost just over the 5% commitment, coming in at $4,441 versus the $4,459 target. This was a big accomplishment because it was a ten-year-old design, and the Gripper team was known for high productivity. They had cut costs significantly over the past two years, and I didn't think we could find much more. We hit target cost at launch, just as the vertical start-up called for."

"John?"

"We saved nearly $10 million in capital expenditure over the original estimate and beat our plan of $8.4 million. Our moonshiners did an amazing job of building a lot of our fixtures and equipment."

Ken jumped in, "Bill, Norm, and Byron did indeed do a phenomenal job, but everyone here also did a great job in simplifying the processes before the equipment build phase. My opinion is that you guys developed a highly flexible factory expansion, with small, right-sized, easy-to-use, and easy-to-maintain equipment."

"Gina?"

"We stress-tested the first 250 units coming off the line outside the factory prior to shipping to the customer and saw ppm levels of 410—less than the 500-ppm target. I believe it'll hold in the field. We hit target quality at launch, also as the vertical start-up called for."

"Gerry?"

"Our start-up lead times are at four weeks as committed, and we've been at 98% on time delivery since launch as well."

"Mike?"

"Our takt time attainment has been pretty high, at 86%, and we needed some overtime to make up for it. But the line is working well. We've been solving small problems with product flow and have shown we can run at rate for periods of time."

"Bill?"

"It's shocking to me, but we started with full standard work for all operators and had them trained beforehand. We were able to do the same with all of the material support folks as well."

"Norm?"

"The volume and size of the product are very similar to our other product lines, yet we're taking up about 50% less space and didn't even use all the space we put into the expansion. We have room left over to expand and grow."

"Byron?"

"I felt we had way more focus on safety through this process than anything I've seen before. We made a big impact on ergonomics, ease of doing the job, stress relief, forklift traffic, and cleanliness and organization of the factory floor. It should lead to much-improved performance in safety and health for our employees."

"Harry?"

"I know that with the simpler equipment that we developed, maintenance will be a lot easier and more effective. It'll cost us much less and ensure a much higher level of uptime than Trail Gripper had experienced at the old plant. Heck, it'll even be better than what we experience on our other product lines in this same factory. It's also a lot easier to use for the operators. We didn't put a lot of bells and whistles on any of the equipment; they were developed just for the exact application they were to do. Yet we made them as flexible as we could for modifications in the future for continuous improvement or product changes."

"Brenda, do you have a benefit you could share with us?"

"The knowledge that my fellow Product Design Engineer Ben and I gained, particularly about Design for Manufacturability concepts, will be invaluable as we design new products in the future."

"Excellent," said Ken.

"And last, but certainly not least, Sylvia."

"I believe that the greatest benefit to all of us and the company as a whole is that we'll all think about improvement differently. I've come to realize that this process required us to think in a totally different way throughout."

"How so?" asked Dave.

"Well, each time we moved onto another aspect of the line to design, whether it was the physical layout or the information and material flows, Ken first defined the concepts he wanted us to base the design on. The Twenty Principles for World-Class Manufacturing is just one example. This made us all think differently about *how* we were to meet our objectives. It helped direct our thought process and our actions in a particular direction. I believe this is one of the more important reasons we were able to meet or exceed the very lofty goals we had."

Everyone nodded in agreement.

Ken smiled and said, "We call that 'target condition' thinking. It's not enough to simply establish numerical goals. We always have to consider the general conditions by which the goals will be met. Although I provided those to you throughout the process, all of you figured out how to make them work.

"What a testimonial to great work. Every single person in this room should be very proud of the work you've done. I'm proud of you, I enjoyed being your coach immensely, and I can tell you that you taught me a lot. I'll miss our daily interactions, but leave here with fond memories."

Pete weighed in to thank Ken for all he did for them and implied that they hadn't seen the last of him. The room erupted into a cacophony of applause.

❋   ❋   ❋

The buses descended on the Hilton Oceanfront Hotel at Hilton Head in South Carolina. Pete had invited the entire 3P team and their spouses for a weekend away together in recognition of one of the best projects he ever had the pleasure to have led. He felt compelled to do something really special for the team, and the trip seemed to be a good way to say "thank you" and celebrate together in a social setting.

Pete stood up after the first evening's meal and gave a heartfelt appreciation speech, talking about the role each person on the team played in the project's success. After the speech, as the applause echoed around the room, Pete said it wasn't much, but as a reward for a job well done, he handed each employee a $2,500 bonus check for their hard work and big results.

❋   ❋   ❋

It was six months later. Dave had just put a picture of his kids atop the filing cabinet when his phone started ringing. He looked out at his assistant, and she smiled in a way that told him he wanted to take that call.

"Mr. Plant Manager," boomed the incoming caller in his familiar voice, addressing Dave by his new title.

"Mr. General Manager. It's been a while," Dave said to his buddy Pete.

Pete had recently received a promotion to general manager, which had precipitated a relocation.

"Have you heard from Ken lately? Enterride is going to hire him to start a Lean journey in a few of our other plants. I have a major new product introduction that I want their organization to help us on. He'll be able to more fully teach us how 3P works with a new product design."

"Didn't you hear? I thought I'd sent you a note; must be in my draft e-mail box. Ken was made a partner at his firm and given an ownership stake. He was really excited when he called me about it. I think he feels he's in the money now."

"He taught me a lot, and I know there's a lot more he can teach me. But you did too. Hiring you was one of the best decisions I ever made. What are you up to today?"

"As far as today goes, Steve Sawyer has another group in here benchmarking what we did with our Trail Gripper transfer and development. He's really been pushing 3P ever since he was here and saw the scale mock-up of the factory."

Pete responded, "Our long-term goal absolutely is to build our own internal capability to fully lead 3P programs on our own, and we'll likely try this on

small ones early on, but I know enough now to realize that we need help growing that capability. So, we're going to continue to rely on Ken's organization to mentor us. Oh, Dave—I have some great news. Steve called me yesterday and told me they're going to invite our entire team to present to the top leaders of the company at the annual leadership conference in January."
"Really? That's fantastic! They're all going to really appreciate this. I can't wait to let everyone know."

Just 11 months after the most successful major product transfer in Enterride's history, and about three years since John Cuso argued forcefully to close the plant, the entire Trail Gripper 3P team looked out from the stage as their CEO Frank Kent told their story. When he finished he asked the roomful of leaders from throughout Enterride to join him in a round of applause to recognize the team for a job well done. It was a truly heartfelt recognition to a team that took some major risks, delivered on stretch commitments, and was wholly deserving of it.

Glossary

3P: *See* Production Preparation Process.

5S: A system designed to organize and standardize a workplace. It consists of five component parts: Sort, Set in order, Shine, Standardize, and Sustain.

7-Ways: A creative thinking technique that requires development of seven different possible methods to address a problem. The technique also allows for a collaboration of ideas and often results in the "best" way being identified.

8 Wastes: Wastes addressed by Lean that include Defects, Overproduction, Waiting, Nonutilized talents, Transportation, Inventory, Motion, and Extra processing. An easy way to remember them is that the first letter of each in this order spells "DOWNTIME."

Batch-and-queue processing: Producing more than one piece of an item and then moving those items forward to the next operation before they are all actually needed there. Thus, these items need to wait in a queue. *Also called* "batch-and-push." *Contrast with* continuous flow.

Bill of Material (BOM): A list of materials and the quantities of each required in the manufacture of a product. A key component of every ERP system.

Continuous flow processing: The process by which items are produced and moved from one processing step to the next, one piece at a time. Each process makes only the one piece that the next process needs, and the transfer batch size is one. Also called "single-piece flow" or "one-piece flow." *Contrast with* batch-and-queue processing.

Cycle time: How frequently an item or product is actually completed by a process, as timed by direct observation. Also, the time it takes an operator to go through all his or her work elements before repeating them.

Enterprise Resource Planning (ERP): A computerized system used to help plan and execute various activities within a company or enterprise. These activities include the calculation of required quantities of purchased and manufactured items and the timing of the replenishment of them, determining the capacity requirements for the production of items, a finance interface to translate operations planning into financial terms, and other functions. Using ERP specifically to schedule production at processes results in push production, because any predetermined schedule is only an estimate of what the next process will actually need.

Fishbone diagram: A visual tool that displays the possible causes of a specific problem. It typically takes the shape of a skeleton of a fish. It is constructed by the continually asking "why?" for each potential cause, adding another "bone" to the "fish" for each response. Also called a Cause & Effect diagram.

Flow: A main objective of the entire Lean production effort, and one of the key concepts that passed directly from Henry Ford to Taiichi Ohno (Toyota's production manager after World War II). Ford recognized that, ideally, production should flow continuously all the way from raw material to the customer and envisioned realizing that ideal through a production system that acted as one long conveyor.

Flow time: *See* lead time.

Inventory turns: A measure of speed in which inventory is converted to sales. It is calculated by dividing the amount of inventory in dollars by the cost of goods sold for a period. The higher the number, the faster the conversion rate. The inverse of inventory turns is the number of days of inventory. Lean enterprise attempts to reduce inventory while maintaining customer service by improving the flow of materials.

Just-in-time: Producing or conveying only the items that are needed by the next process when they are needed and in the quantity needed.

Kaizen: Continuously improving in incremental steps. Typically refers to improvement at the process level.

Kanban: A signaling device that gives instruction for production or withdrawal (conveyance) of items in a pull system. The signal can take various forms (e.g., card, container).

Lead time: The time required for one piece to move all the way through a process or value stream, from start to finish. Envision timing a marked item as it moves from beginning to end. Can also be calculated by multiplying takt time by standard work-in-process (SWIP).

Lean: A systematic approach to identifying and eliminating waste (non-value-added activities) through continuous improvement by flowing the product at the pull of the customer in pursuit of perfection.

Lean enterprise: The organization that fully understands, communicates, implements, and sustains Lean concepts seamlessly throughout all operational and functional areas.

Line balancing: A process in which work elements are evenly distributed within the value stream to meet takt. An Operator Balance chart is often used to help achieve this.

Material handlers: Production-support persons who travel repeatedly along scheduled routes within a facility to transfer materials, supplies, and parts in response to pull signals, and to make paced withdrawals of finished goods at pacemaker processes.

Milk run: Routing a delivery vehicle in a way that allows it to make pick-ups or drop-offs at multiple locations on a single travel loop, as opposed to making separate trips to each location.

Mistake-proofing: Refers to methods that help operators avoid mistakes in their work. A common example is a product design with physical attributes that make it impossible to assemble incorrectly. *Also called* poka-yoke.

Mixed-model scheduling: An approach to scheduling final production processes that smooths out demand on the supply chain by producing some of each item over the shortest possible time horizon.

Moonshiner: A very mechanically oriented person who is creative and skilled in creating simple low-cost tools and equipment.

Moonshining: The practice of developing simple, low-cost tool and equipment solutions to problems that operators are having or wastes that exist in the process.

Non-Value-Added: Any activity that does not add market form or function, or is not necessary. (These activities should be eliminated, simplified, or reduced.)

Operator balance chart: A graphic tool that assists in the creation of continuous flow in a multi-step, multi-operation process by distributing work elements in an even way between operators, and in relation to takt time. *See also* Yamazumi board.

Overall Equipment Effectiveness (OEE): A measure of equipment availability that consists of several components: speed losses, minor stoppages and adjustments, defects, unplanned downtime, set-up, or changeover.

Parts per million (ppm): A measure of quality that is calculated by projecting a defect rate typically expressed as a percentage over a million units. For example, a 5 percent defect rate would be equivalent to 50,000 ppm.

Plan-for-Every-Part (PFEP): A detailed plan for each part used in a production process. The plan will include the part number, the amount used daily, the location(s) of use, the location of storage, order frequency, container size, type, and weight, and other relevant information.

Point-of-Use-Storage (POUS): Raw material stored at the workstation where it is used.

Process-at-a-Glance (PAAG): A form that is used to envision the flow of materials through several steps in a sequence and to conceptualize possible methods for processing them. It includes possible methods, tools, fixtures, machines, and other items. Is often used during a 7-Ways exercise.

Process kaizen: Improvements made in an individual process or in a specific area. Sometimes called "point kaizen."

Processing time: The time a product is actually being worked on in a machine or work area.

Production Preparation Process (3P): A disciplined method for designing a Lean production process for a new product or for fundamentally redesigning the production process for an existing product. In an ideal situation, the product and process are designed simultaneously.

Pull system: An information system for controlling and improving the flow of materials and information, and for allocating resources based on actual consumption, not forecasted demand.

Push system: A system where resources are provided to the consumer based on forecasts or schedules rather than actual demand.

Standard Work Combination sheet: A form that shows the combination of manual work time, walk time, and machine processing time for each operation in a production sequence. The combination of work can vary based on the takt time.

Standard Work-in-Process (SWIP): A set amount of desired and allowable inventory within a process. Inventory beyond this amount indicates a nonstandard condition that must be addressed.

Standardized work: Establishing precise procedures for each operator's work in a production process based on three elements: takt time, work sequence, and standard inventory. Several basis forms are used, including the Standard Work Combination sheet.

System kaizen: Improvement aimed at an entire value system. Also called kaikaku, which means radical or revolutionary improvement.

Takt time: The rate of customer demand: how often the customer requires one finished item. Takt time is used to design assembly and pacemaker processes, to assess production conditions, to develop material handling containerization and routes, to determine problem-response requirements, and so on. Takt is the heartbeat of a Lean system. Takt time is calculated by dividing effective working time by the quantity the customer requires in that time.

Total Productive Maintenance (TPM): A systematic approach to the elimination of equipment downtime as a waste factor.

Trystorming: Hands-on experimentation that puts to the test an idea in order to assess its effectiveness.

Value added: Any activity that increases the market form or function of the product or service. (These are things the customer is willing to pay for.)

Value stream: All activities, both value added and non-value-added, required to bring a product from raw material into the hands of the customer, a customer requirement from order to delivery, and a design from concept to launch. Value stream improvement usually begins at the door-to-door level within a facility and then expands outward to eventually encompass the full value stream.

Value stream mapping: A pencil-and-paper tool used in the following two stages: (1) To follow a product's production path from beginning to end and draw a visual representation of every process in the material and information flows and (2) to then draw a future state map of how value should flow.

Waste: Any activity that consumes resources but creates no value for the customer.

Water spider: Another term for material handler. A person who moves materials through a production process within a facility. In a Lean production system, water spiders do much more than just deliver materials. They also convey information (e.g., kanbans) that is critical in controlling the flow of materials.

Yamazumi board: A stack chart that is used to visibly display process times so that steps can be combined to achieve balance. *See also* Operator Balance Chart.

Index

Page numbers followed by e indicate exhibit
Page numbers followed by t indicate table

Printed in the United States
by Baker & Taylor Publisher Services